Earth
Anchors

TITLES IN THE SERIES

Architectural Acoustics
M. David Egan
ISBN 13: 978-1-932159-78-3, ISBN 10: 1-932159-78-9, 448 pages

Earth Anchors
By Braja M. Das
ISBN 13: 978-1-932159-72-1, ISBN 10: 1-932159-72-X, 242 pages

Limit Analysis and Soil Plasticity
By Wai-Fah Chen
ISBN 13: 978-1-932159-73-8, ISBN 10: 1-932159-73-8, 638 pages

Plasticity in Reinforced Concrete
By Wai-Fah Chen
ISBN 13: 978-1-932159-74-5, ISBN 10: 1-932159-74-6, 474 pages

Plasticity for Structural Engineers
By Wai-Fah Chen & Da-Jian Han
ISBN 13: 978-1-932159-75-2, ISBN 10: 1-932159-75-4, 606 pages

Theoretical Foundation Engineering
By Braja M. Das
ISBN 13: 978-1-932159-71-4, ISBN 10: 1-932159-71-1, 440 pages

Theory of Beam-Columns, Volume 1: Space Behavior and Design
By Wai-Fah Chen & Toshio Atsuta
ISBN 13: 978-1-932159-77-6, ISBN 10: 1-932159-77-0, 732 pages

Theory of Beam-Columns, Volume 2: In-Plane Behavior and Design
By Wai-Fah Chen & Toshio Atsuta
ISBN 13: 978-1-932159-76-2, ISBN 10: 1-932159-76-9, 513 pages

Earth Anchors

by Braja M. Das, Ph.D.
Henderson, Nevada

J.ROSS PUBLISHING

Copyright ©2007 by Braja M. Das

ISBN-10: 1-932159-72-X
ISBN-13: 978-1-932159-72-1

Printed and bound in the U.S.A. Printed on acid-free paper
10 9 8 7 6 5 4 3 2 1

This J. Ross Publishing edition, first published in 2007, is an unabridged republication of the work originally published by Elsevier Science Publishers B.V., in 1990.

Library of Congress Cataloging-in-Publication Data

Das, Braja M., 1941–
    Earth anchors / by Braja M. Das.
      p. cm.
    Reprint. Originally published: Amsterdam : Elsevier Science, c1990.
    Includes bibliographical references and index.
    ISBN-10: 1-932159-72-X (pbk : alk. paper)
    ISBN-13: 978-1-932159-72-1 (pbk : alk. paper)
    1. Foundations. 2. Anchorage (Structural engineering) I. Title.
    TA775.D226 2007
    624.1'5—dc22   2006101134

Phone: (954) 727-9333
Fax: (561) 892-0700
Web: www.jrosspub.com

To
My wife Janice
and
daughter Valerie

PREFACE

Anchors are primarily used in construction of foundations of earth-supported and earth-retaining structures. The fundamental reason for using earth anchors in construction is to transmit the outwardly-directed load to the soil at a greater depth and/or farther away from the structure. Although earth anchors have been used in practice for several hundred years, proper theoretical developments for purposes of modern engineering design have taken place only during the past twenty years or so. This book summarizes most theoretical and experimental works directed toward the development of proper relationships for ultimate and allowable holding capacity of earth anchors.

The book contains six chapters with detailed discussions on horizontal, vertical and inclined plate anchors, helical anchors, and anchor piles. The discussions have been limited to the failure mechanisms in the soil and procedures to calculate the ultimate and allowable loads. No attempt has been made to describe the construction procedures for the installation of the anchors. Modifications to the contents of the book will, of course, be necessary with future developments and changes in the state-of-the-art during the preparation of the second edition of this book. I sincerely hope this book will be helpful to designers and researchers in the area of earth anchors.

I am grateful to my wife, Janice, for typing the entire manuscript in camera-ready form. She also drew several figures and developed a number of tables My colleague, Dr. Bruce A. DeVantier of the Department of Civil Engineering and Mechanics at Southern Illinois University at Carbondale, also helped prepare some of the graphs and tables. Thanks are also due to Robert L. Goodman, Director of Editorial

Research-North America for Elsevier Scientific Publishing Company and Jacques Kiebert of the Earth Sciences Division of Elsevier Science Publishers B.V. who were most instrumental in the initiation of the manuscript.

— **Braja M. Das**
Carbondale
May, 1990

CONTENTS

Chapter 1
ANCHORS - GENERAL

1.1 INTRODUCTION

Anchors are primarily designed and constructed to resist outwardly- directed loads imposed on the foundation of a structure. These outwardly-directed loads are transmitted to the soil at a greater depth by the anchors.

Buried anchors have been used for thousands of years to stabilize structures. Tents are the oldest structures which were stabilized by using anchors or stakes. Until the middle of the nineteenth century, anchors were primarily used for stabilizing fairly lightweight structures. With the design and construction of large suspension bridges, very large loads were transmitted to the bridge foundation. In order to support these loads, permanent anchoring systems in rock medium were gradually developed and constructed.

With the development and construction of special lightweight structures such as lattice transmission towers and radar towers, design of special tension anchoring systems for foundations became necessary, primarily because the wind load created reactions which were greater than the self-weight of the structures.

Earth anchors of various types are now used for uplift resistance of transmission towers, utility poles, aircraft moorings, submerged pipelines, and tunnels. Anchors are also used for tieback resistance of earth-retaining structures, waterfront structures, at bends in pressure pipelines, and when it is necessary to control thermal stress.

The earlier forms of anchors used in soil for resisting vertically-directed uplifting loads were *screw anchors*. Figure 1-1 shows two different configurations of screw anchors. These anchors were simply twisted into the ground up to a pre-estimated depth, and then tied to the foundation. They were either used singly or in groups.

In general, at the present time, anchors placed in soil can be divided into five basic categories: plate anchors, direct embedment anchors, helical anchors, grouted anchors, and anchor piles and drilled shafts. Some authors refer to plate anchors as direct embedment anchors.

2

1.2 PLATE ANCHORS

Plate anchors may be made of steel plates, precast concrete slabs, poured concrete slabs, timber sheets, and so forth. They may be horizontal to resist
vertically-directed uplifting load, inclined to resist axial pullout load, or
vertical to resist horizontally-directed pullout load, as shown in Figures 1-
2a, 1-2b, and 1-2c. These anchors can be installed by excavating the ground
to the required depth and then backfilling and compacting with good quality
soil. They may be referred to as *backfilled plate anchors* (Figure 1-3a). In
many cases, plate anchors may be installed in excavated trenches as shown in
Figure 1-3b. These anchors are then attached to tie rods which may either be
driven or placed through augured holes. Anchors placed in this way are
referred to as *direct bearing plate anchors*. In the construction of sheet
pile walls, primarily used for waterfront structures, vertical backfilled or
direct bearing plate anchors are common. Figure 1-4a shows the cross section
of a sheet pile wall with vertical anchor. The vertical anchors of height h
and width B and spaced with a center-to-center spacing of S are tied to the
sheet pile wall as shown in Figure 1-4b.

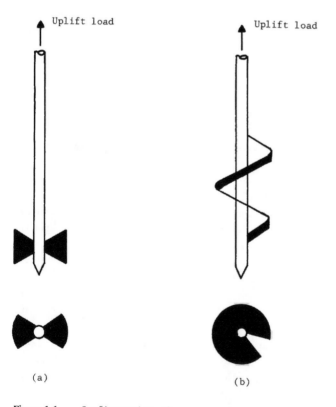

(a)   (b)

Figure 1-1   Configuration of screw anchors

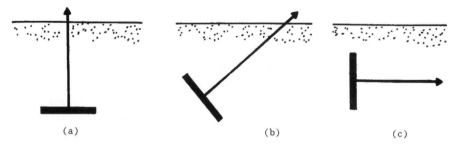

Figure 1-2 Plate anchors: (a) horizontal plate anchor; (b) inclined plate anchor; (c) vertical plate anchor

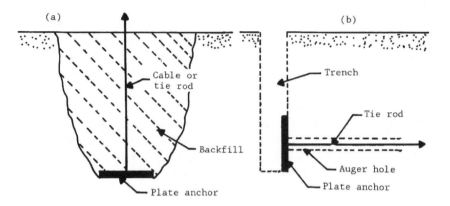

Figure 1-3 Installation of plate anchors: (a) backfilled plate anchor; (b) direct bearing plate anchor

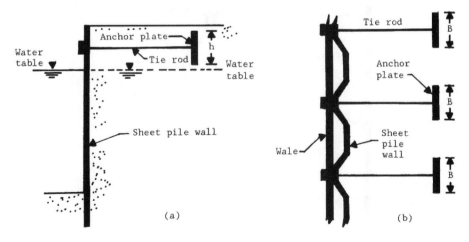

Figure 1-4 Use of vertical plate anchor in sheet pile wall: (a) section; (b) plan

4

In many cases, horizontal anchor beams along with batter piles can also be used in the construction of sheet pile walls (Figure 1-5).

1.3   DIRECT EMBEDMENT ANCHORS

Direct embedment anchors are similar in nature to direct bearing plate anchors (Figure 1-6). They may be triangular or take any other penetrative shape, and they are installed vertically by driving with a rod to a desired depth. After the desired depth is reached, the rod is withdrawn and the cable is tensioned to rotate the anchor through an angle of 90° into its final position.

1.4   HELICAL ANCHORS

Helical anchors consist of a steel shaft with one or more helices attached to it (Figure 1-7). For multi-helix anchors, the pitch and center to center spacing of the helices can be varied so that the upper helices follow the lower ones. This helps reduce the disturbance in the soil. Figures 1-8 and 1-9 show photographs of helical anchors with one and two helices. The schematic diagram and the photograph of the installation of a helical anchor are shown in Figures 1-10 and 1-11, respectively. These anchors are driven into the ground in a rotating manner using truck- or trailer-mounted auguring equipment where the soil conditions permit. An axial load is applied to the shaft while rotating to advance it into the ground. While installing these augers in soils mixed with gravel and large boulders, care should be taken to avoid possible damage to the helices.

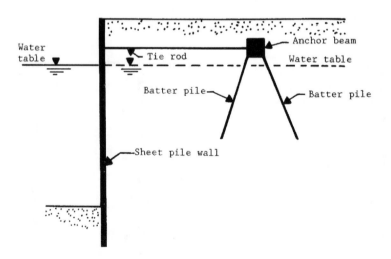

Figure 1-5   Use of horizontal anchor beam with batter piles in sheet pile wall

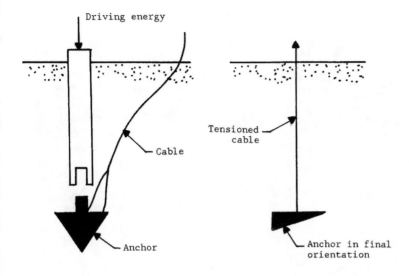

Figure 1-6   Direct embedment anchor (redrawn after Kulhawy, 1985)

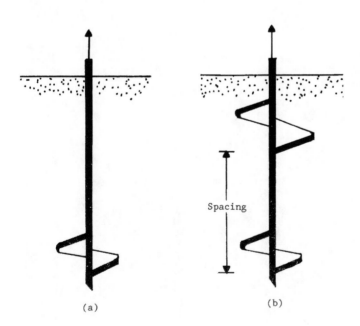

Figure 1-7   Helical anchors

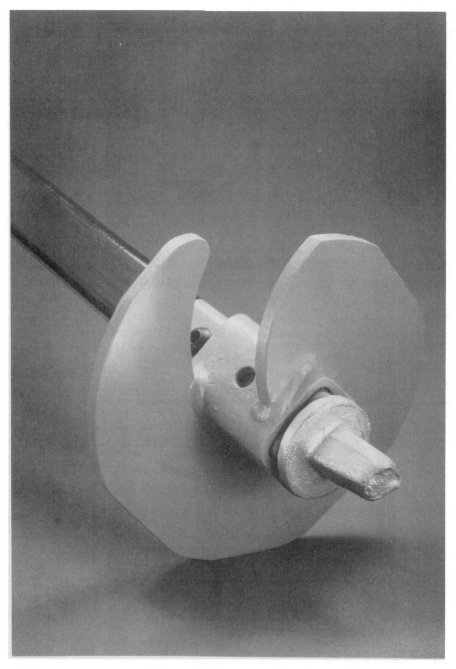

Figure 1-8   Helical anchors with one helix (courtesy of A.B. Chance Co., Centralia, Missouri, USA)

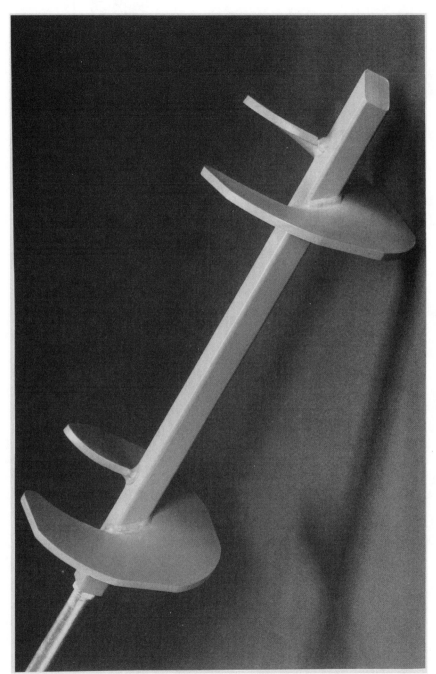

Figure 1-9   Helical anchor with two helices (courtesy of A. B. Chance Co., Centralia, Missouri, USA)

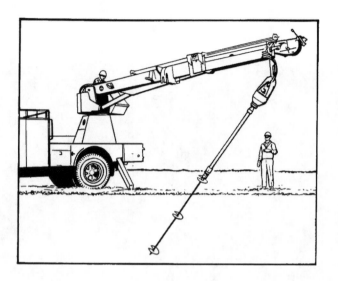

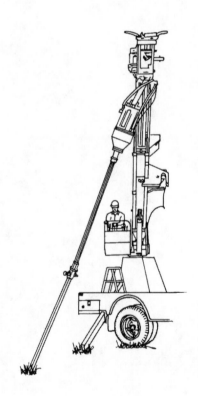

Figure 1-10   Installation of helical anchor

Figure 1-11 Photograph of installation of helical anchor (courtesy of A. B. Chance Co., Centralia, Missouri, USA)

10

Helical anchors can resist tensile loads on the foundation; however, at the same time, they can also supply additional bearing capacity to the foundation (under downward-loading condition) developed at the helix-soil interface.

Helical anchors are becoming increasingly popular in the construction of electric transmission tower foundations in the United States. They may be installed in either a vertical or an inclined position.

1.5 GROUTED ANCHORS

Grouted anchors primarily consist of placing a steel bar or steel cable into a pre-drilled hole and then filling the hole with cement grout. Figure 1-12 shows various types of grouted anchors, brief explanations of which are given below.

1.   *Gravity Type.* For this type of anchor, the grout is poured into the hole from the ground surface without any pressure (Figure 1-12a).

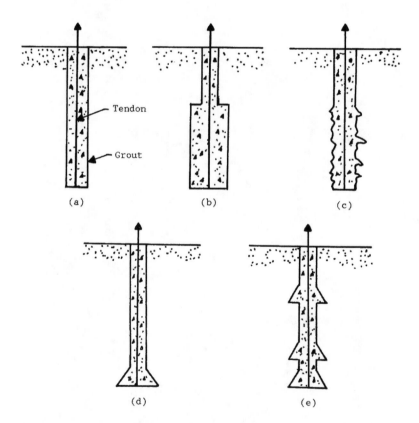

Figure 1-12   Grouted anchors: (a) gravity; (b) low pressure; (c) high pressure; (d) single bell; (e) multiple bell (redrawn after Kulhawy, 1985)

2. *Low Pressure Type*. For this type of anchor, the grout is injected into the hole at pressures up to the overburden pressure (Figure 1-12b).

3. *High Pressure Type*. For anchors of this type, the grout is injected at high pressure. This pressure increases the effective diameter of the anchor and compacts the loose soil around it (Figure 1-12c).

4. *Single and Multi-bell Type*. This is primarily a gravity type anchor; however, single or multiple bells are made in the ground mechanically before grouting (Figures 1-12d and 1-12e).

Grouted anchors can be used in many construction projects, such as sheet pile walls (Figure 1-13a), revetment of rock retaining walls (Figure 1-13b), basement floors to resist buoyancy (Figure 1-13c), and foundations of transmission towers to resist overturning.

1.6 ANCHOR PILES AND DRILLED SHAFTS

Piles and drilled shafts (Figure 1-14) can be used in the construction of foundations subjected to uplift where soil conditions are poor, or for very heavily loaded foundations. They serve dual purposes: that is, they help support the downward load on the foundation of the structure, and they also resist uplift.

1.7 COVERAGE OF THE TEXT

During the last fifteen to twenty years, experimental and analytical research relating to the holding capacity of various types of anchors have accelerated,

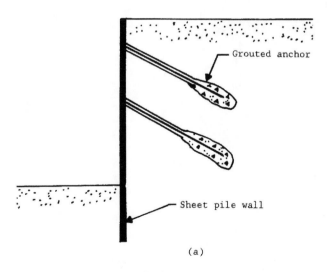

(a)

Figure 1-13 (Continued)

12

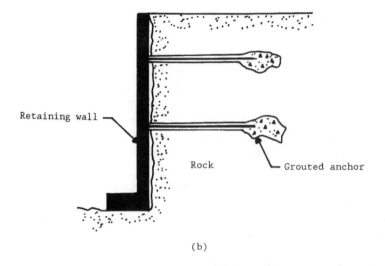

Retaining wall

Rock

Grouted anchor

(b)

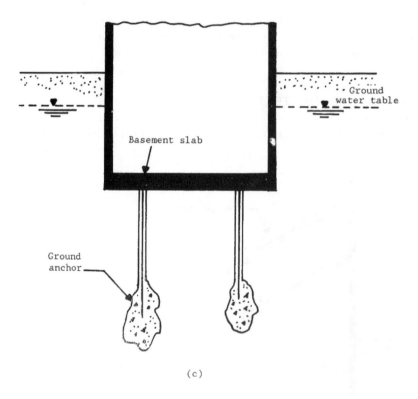

Ground
water table

Basement slab

Ground
anchor

(c)

Figure 1-13   Use of grouted anchors in (a) sheet pile wall; (b)
revetment of rock retaining wall; (c) floor of
basement

13

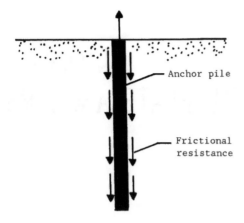

Figure 1-14   Anchor pile and drilled shaft subjected to uplifting
load

and results of those works have been published in various journals and techni-
cal conference proceedings.   The purpose of this text is to present in a
systematic manner a comprehensive review of some of the recent studies.   The
subjects covered are for evaluation of the holding capacities of *plate anchors
oriented in horizontal, inclined, and vertical manner* in soil, *helical anchors*, and *piles subjected to vertical uplift*.   No attempt has been made to
provide either the details for the *placement* of the anchors in the field or
the *construction techniques*.   Valuable information in these areas can be obtained from the work of Hanna (1982) and others.   No aspects of grouted
anchors are covered in this text, since valuable information is available from
several other well-organized sources (Hanna, 1982; Littlejohn, 1970).   In
spite of the accelerated pace of research work on various aspects of anchors
at the present time, adequate field verification is often lacking in several
instances.   These shortcomings will also be outlined in the text.

REFERENCES

Hanna, T.H., 1982.   *Foundations in tension-ground anchor*   Trans Tech
    Publications and McGraw Hill Book Company.
Kulhawy, F.H., 1985.   Uplift behavior of shallow soil anchors--an overview.
    Proc., Uplift Behavior of Anchor Foundations in Soil, ASCE, 1-25.
Littlejohn, G.S., 1970.   Soil anchors.   *Proc.*, Conf. Ground Engg., London, 33-
    44.

Chapter 2
HORIZONTAL PLATE ANCHORS

2.1 INTRODUCTION

As briefly discussed in Chapter 1, horizontal plate anchors are used in the construction of foundations subjected to uplifting load. During the last thirty years or so, a number of increasingly sophisticated theories have been developed to predict the ultimate uplift capacity of horizontal anchors embedded in various types of soils. In this chapter the development of those theories will be discussed. The chapter has been divided into two major parts: (a) plate anchors in sand and (b) plate anchors in clay.

Figure 2-1 shows a plate anchor having a width h and a length B ($B \geq h$). The embedment depth of this plate anchor is H measured from the ground surface. The *embedment ratio* is defined as the ratio of the depth of embedment to the width of the anchor, (that is, H/h). If such an anchor is placed at a relatively shallow depth (that is, small embedment ratio), at ultimate load the failure surface will extend to the ground surface (Figure 2-2). The angle

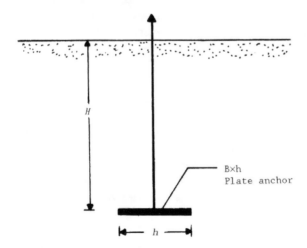

$B \times h$
Plate anchor

H

h

Figure 2-1　　Geometric parameters of a horizontal plate anchor

α at which the failure surface intersects the horizontal ground surface will vary. For loose sand and soft clayey soils, α may be equal to 90°; however, for dense sand and stiff clays, this angle may be close to 45 -$\phi/2$ (where ϕ = soil friction angle). This type of behavior of an anchor is referred to as *shallow anchor condition*. If the anchor is located at a relatively large embedment ratio, the failure surface in soil at ultimate load does not extend to the ground surface (that is, local shear failure in soil located around the anchor takes place). This is referred to as *deep anchor condition*.

For a given anchor, the gross ultimate uplift capacity can be defined as

$$Q_{u(g)} = Q_u + W_a \qquad (2-1)$$

where $Q_{u(g)}$ = gross ultimate uplift capacity
 Q_u = net ultimate uplift capacity
 W_a = effective self-weight of the anchor

The net ultimate uplift capacity is the sum of the effective weight of the soil located in the failure zone and the shearing resistance developed along the failure surface.

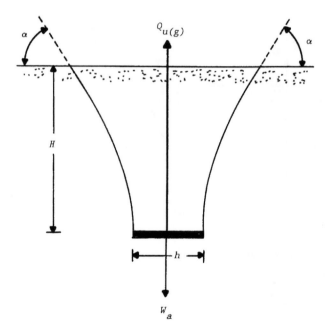

Figure 2-2   Shallow horizontal anchor

16

2.2 EARLY THEORIES

Soil Cone Method

Some of the early theories to determine the net ultimate uplift capacity Q_u were restricted to *shallow circular* plate anchors. Mors (1959) proposed that the failure surface in soil at ultimate load may be approximated as a truncated cone having an apex angle of $\theta = 90° + \phi/2$ as shown in Figure 2-3. The net ultimate uplift capacity may be assumed to be equal to the weight of the soil located inside the failure surface. Thus

$$Q_u = \gamma V \tag{2-2}$$

where V = volume of soil in the truncated cone
γ = unit weight of soil

However

$$V = \frac{\pi}{3}H\left\{h^2 + [h + 2H \cot(45 \quad \frac{\phi}{4})]^2 + h[h + 2H \cot(45 \quad \frac{\phi}{4})]\right\}$$

or

$$V = \frac{\pi H}{3}\left[3h^2 + 4H^2\cot^2(45 \quad \frac{\phi}{4}) + 6Hh \cot(45 \quad \frac{\phi}{4})\right] \tag{2-3}$$

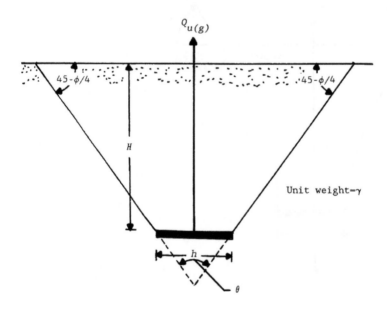

Figure 2-3 Mors' theory--soil cone method (*Note*: $\theta = 90 + \phi/2$; h = diameter of anchor plate)

It needs to be pointed out that the shearing resistance developed along the failure surface has been neglected in Equation (2-2).

A similar theory was also proposed by Downs and Chieurzzi (1966),who suggested that the apex angle θ be taken as being equal to 60° as shown in Figure 2-4. For this case

$$Q_u = \gamma V = \frac{\pi \gamma H^3}{3}\left\{ h^2 + [h + 2H \cot 60°]^2 + h(h + 2H \cot 60°)\right\}$$

$$= \frac{\pi \gamma H^3}{3}(3h^2 + 1.333H^2 + 3.464H) \tag{2-4}$$

Friction Cylinder Method

In many cases in the past, the friction cylinder method was used to estimate the uplift capacity of *shallow circular* anchor plates. In this type of calculation, the friction surface in the soil was assumed to be cylindrical as shown in Figure 2-5a. For cohesionless soils, the net ultimate load was taken as the sum of the weight of the soil located inside the failure cylinder and the frictional resistance derived along the failure surface. Thus

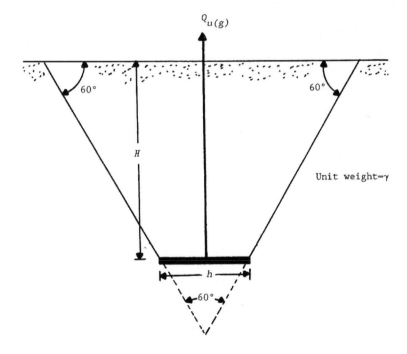

Figure 2-4   Downs and Chieurzzi's theory--soil cone method
(*Note*: h = diameter of anchor plate)

18

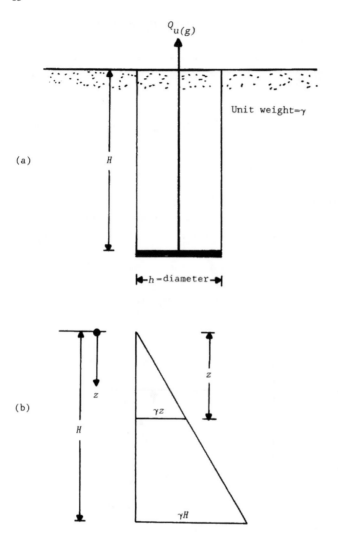

Figure 2-5   Friction cylinder method

$$Q_u = (\frac{\pi h^2}{4})(H)(\gamma) + \int_0^H (\sigma_o' \tan \phi)dz$$

where σ_o' = effective overburden pressure at a depth z measured from the ground
surface (Figure 2-5b)

ϕ = soil friction angle

So

$$Q_u = \frac{\pi H h^2 \gamma}{4} + \int_0^H (\pi h)(\gamma z \tan \phi)dz = \frac{\pi H h^2 \gamma}{4} + (\frac{\pi h H^2 \gamma}{2})(\tan \phi) \qquad (2-5)$$

In a similar manner, for saturated cohesive soil

$$Q_u = \frac{\pi H h^2 \gamma}{4} \quad + \quad (\pi H h) \quad (c_u) \qquad (2-6)$$
$$\uparrow$$
surface area of the
cylindrical failure
surface

where c_u = undrained cohesion

Ireland (1963) proposed the following relationships for shallow anchors embedded in sand as well as silts and clay.

$$Q_u = \frac{\pi H h^2 \gamma}{4} + \frac{\pi}{2}\gamma h H^2 K_o \tan \phi \qquad (2-7)$$

where K_o = coefficient of lateral earth pressure

Ireland (1963) also recommended the following values for K_o and ϕ.

$$K_o = \begin{cases} 0.5 \text{ for granular soil} \\ 0.4 \text{ for silts and clays} \end{cases}$$

$$\phi = \begin{cases} 30° \text{ for granular soil} \\ 20° \text{ for silts and clays} \end{cases}$$

RECENT SOLUTIONS FOR ANCHORS IN SAND

2.3   NET ULTIMATE UPLIFT CAPACITY

More recently, several theoretical and semi-empirical methods were developed to predict the ultimate uplifting load of strip, circular and rectangular anchors embedded in sand.   Some of these theories are briefly described in this section.

2.3.1   Balla's Theory

Based on several model and field test results in dense soil, Balla (1961) established that, for *shallow circular anchors*, the failure surface in soil will be as shown in Figure 2-6.   Note from the figure that *aa'* and *bb'* are arcs of a circle.   The angle α is equal to $45 \quad \phi/2$.   The radius of the circle, of which *aa'* and *bb'* are arcs, is equal to

$$r = \frac{H}{\sin(45 + \frac{\phi}{2})} \qquad (2-8)$$

The net ultimate uplift capacity of the anchor is the sum of two components: (a) weight of the soil in the failure zone and (b) the shearing resistance developed along the failure surface. Thus

$$Q_u = H^3 \gamma [F_1(\phi, \frac{H}{h}) + F_3(\phi, \frac{H}{h})] \qquad (2-9)$$

The sums of the functions $F_1(\phi, H/h)$ and $F_3(\phi, H/h)$ developed by Balla (1961) are plotted in Figure 2-7 for various values of the soil friction angle ϕ and embedment ratio H/h. The general nature of the plot of Q_u versus H/h will be like that in Figure 2-8.

In general, Balla's theory is in good agreement for the uplift capacity of anchors embedded in dense sand at an embedment ratio of $H/h \leq 5$. However, for anchors located in loose and medium sand, the theory overestimates the net ultimate uplift capacity. The main reason that Balla's theory overestimates the net ultimate uplift capacity for $H/h >$ about 5 even in dense sand is because it is essentially deep anchor condition, and the failure surface does not extend to the ground surface.

The simplest procedure to determine the embedment ratio at which deep anchor condition is reached may be determined by plotting the nondimensional breakout factor F_q against H/h as shown in Figure 2-9.

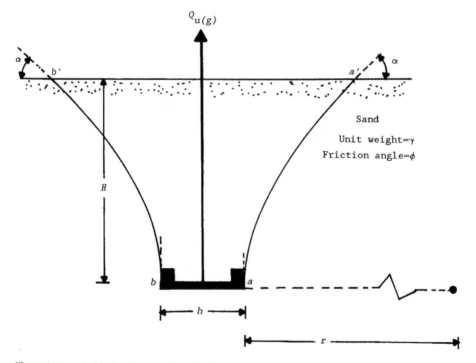

Figure 2-6   Balla's theory for shallow circular anchor plate

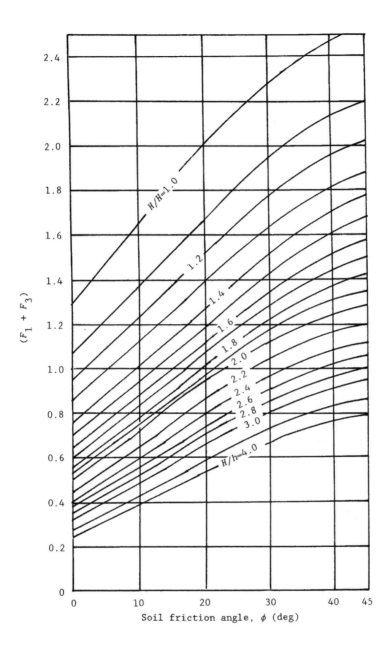

Figure 2-7 Variation of $F_1 + F_3$ based on Balla's theory (1961)

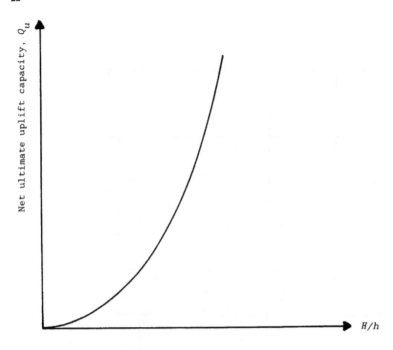

Figure 2-8   Nature of variation of Q_u with H/h

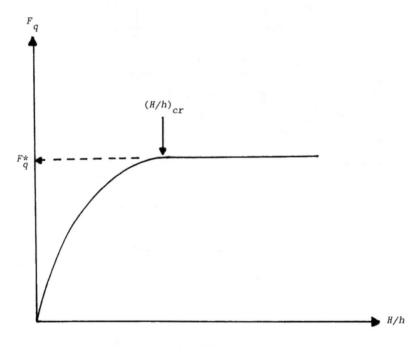

Figure 2-9   Nature of variation of F_q with H/h

The breakout factor is defined as

$$F_q = \frac{Q_u}{\gamma AH} \tag{2-10}$$

where A = area of the anchor plate

The breakout factor increases with H/h up to a maximum value of $F_q = F_q^*$ at H/h = $(H/h)_{cr}$. For $H/h > (H/h)_{cr}$ the breakout factor remains practically constant (that is, F_q^*). Anchors located at an embedment ratio of $H/h \le (H/h)_{cr}$ are shallow anchors, and those located at $H/h > (H/h)_{cr}$ are deep anchors.

2.3.2 Baker and Kondner's Empirical Relationship

Baker and Kondner (1966) conducted several laboratory model tests and, by using dimensional analysis, they proposed the following relationships.

$$Q_u = C_1 H h^2 \gamma + C_2 H^3 \gamma \quad \text{(for shallow circular anchors)} \tag{2-11}$$

$$Q_u = 170 h^3 \gamma + C_3 h^2 t \gamma + C_4 H h + \gamma \quad \text{(for deep circular anchors)} \tag{2-12}$$

where $\quad t$ = thickness of the anchor plate

C_1, C_2, C_3, C_4 = constants which are functions of the soil friction angle and the relative density of compaction

For shallow anchors the model test results of Baker and Kondner agreed well with the theory of Balla (1961). Those tests were conducted in a dense sand with $\phi = 42°$

2.3.3 Mariupol'skii's Theory

Mariupol'skii (1965) proposed separate mathematical formulations for estimation of the ultimate uplift capacity of shallow and deep circular anchors. According to his theory, for shallow anchors the progressive failure mechanism commences with compression of the soil located above the anchor plate (Figure 2-10). This compression occurs within a column of soil the same diameter as the anchor plate. Hence, the initial force consists of the following components:

a. the effective weight of the anchor,

b. the effective weight of the soil column of diameter h and height H, and

c. the friction and cohesion along the surface of the soil column.

As pullout progresses there is continued compaction of soil, and this leads to an increase in the vertical compressive stress. Thus there is a continued increase in the frictional resistance along the surface of the soil column. The

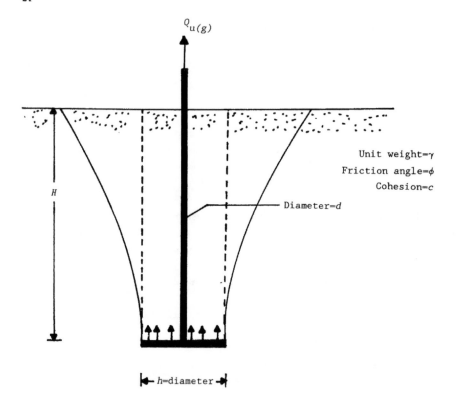

$Q_{u(g)}$

Unit weight=γ
Friction angle=ϕ
Cohesion=c

Diameter=d

H

|← h=diameter →|

Figure 2-10   Mariupol'skii's theory for shallow circular plate anchor

increase of the frictional resistance entrains adjacent rings of soil. Ultimately sufficient tensile stress is developed so that failure occurs with the separation of soil in the form of a cone with a curvilinear geneatrix. The net ultimate uplift capacity thus calculated by this theory can be given as

$$Q_u = \frac{\pi}{4}(h^2 - d^2) \left\{ \frac{\gamma H[1 - (\frac{d}{h})^2 + 2K_o(\frac{H}{h})\tan\phi] + 4c(\frac{H}{h})]}{1 - (\frac{H}{h})^2 - 2n(\frac{H}{h})} \right\} \qquad (2-13)$$

where K_o = lateral earth pressure coefficient
c = cohesion
n = an empirical coefficient ≈ 0.025ϕ (degrees)
d = diameter of anchor shaft

For sand, $c = 0$. So

$$Q_u = \frac{\pi}{4}(h^2 \quad d^2)\left\{\frac{\gamma H[1 \quad (\frac{d}{h})^2 + 2K_o(\frac{H}{h})\tan\phi]}{1 \quad (\frac{H}{h})^2 \quad 2n(\frac{H}{h})}\right\} \qquad (2\text{-}14)$$

For deep anchors, it was assumed that under the applied load the anchor will reach a limiting condition, after which additional work is required to raise the anchor through a distance L which is equivalent to the work required to expand a cylindrical cavity of height L and diameter d to a diameter h as shown in Figure 2-11. Based on this concept, the net ultimate uplift capacity can be expressed as

$$Q_u = (\frac{\pi q_o}{2})[\frac{(h^2 - d^2)}{2 \quad \tan\phi}] + f(\pi d)[H \quad (h \quad d)] \qquad (2\text{-}15)$$

↑
effective length
of anchor stem

where q_o = radial pressure under which the cavity is expanded

f = unit skin resistance along the stem of the anchor

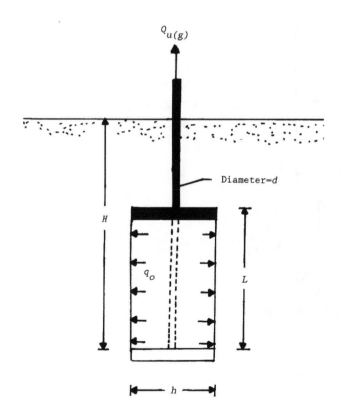

Figure 2-11   Mariupol'skii's theory for deep circular plate anchor

It was recommended that the lower of the two values [that is, those calculated from either Equation (2-14) or (2-15)] be adopted for design. This was primarily because the limit of $H/h = (H/h)_{cr}$ for deep anchor condition was not clearly established.

2.3.4 Meyerhof and Adams' Theory

Meyerhof and Adams (1968) proposed a semitheoretical relationship for estimation of the ultimate uplift capacity of *strip, rectangular* and *circular* anchors. It needs to be pointed out that this is the only theory presently available for estimation of Q_u for rectangular anchors. The principles of this theory can be explained by considering a *shallow strip* anchor embedded in sand as shown in Figure 2-12.

At ultimate load the failure surface in soil makes an angle α with the horizontal. The magnitude of α depends on several factors such as the relative density of compaction and the angle of friction of the soil, and it varies between 90° $\phi/3$ to 90° $2\phi/3$ with an average of about 90 $\phi/2$. Let

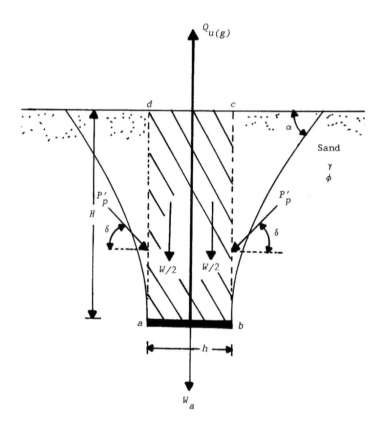

Figure 2-12   Derivation of Equation (2-20)

us consider the free body diagram of the soil located in the zone *abcd*. For stability consideration, the following forces *per unit length of the anchor* need to be considered:

a.   the weight of the soil, W, and

b.   the passive force P_p' per unit length along the faces *ab* and *cd*. The force P_p' is inclined at an angle δ to the horizontal. For an average value of $\alpha = 90 \quad \phi/2$, the magnitude of δ is about $(2/3)\phi$.

Note that

$$W = \gamma H h \qquad (2\text{-}16)$$

$$P_p' = \frac{P_h'}{\cos \delta} = (\frac{1}{2})(\frac{1}{\cos \delta})(K_{ph}\gamma H^2) \qquad (2\text{-}17)$$

where P_h' = horizontal component of the passive force

K_{ph} = horizontal component of the passive earth pressure coefficient

Now, for equilibrium, summing the vertical components of all forces

$$\Sigma F_v = 0$$

$$Q_{u(g)} = W + 2P_p'\sin \delta + W_a$$

$$Q_{u(g)} \quad W_a = W + 2(P_p'\cos \delta)\tan \delta$$

$$Q_u = W + 2P_h'\tan \delta$$

or

$$Q_u = W + 2(\frac{1}{2}K_{ph}\gamma H^2)\tan \phi = W + K_{ph}\gamma H^2 \tan \delta \qquad (2\text{-}18)$$

The passive earth pressure coefficient based on the curved failure surface for $\delta \approx (2/3)\phi$ can be obtained from Caquot and Kerisel (1949) Furthermore, it is convenient to express K_{ph} tan δ in the form

$$K_u \tan \phi = K_{ph}\tan \delta \qquad (2\text{-}19)$$

Combining Equations (2-18) and (2-19) we obtain

$$Q_u = W + K_u\gamma H^2 \tan \phi \qquad (2\text{-}20)$$

where K_u = nominal uplift coefficient

28

The variation of the nominal uplift coefficient K_u with the soil friction angle ϕ is shown in Figure 2-13. It falls within a narrow range and may be taken as equal to 0.95 for all values of ϕ varying from 30° to about 48°

As discussed in Section 2.3.1, the nondimensional breakout factor is defined as

$$F_q = \frac{Q_u}{\gamma A H}$$

For strip anchors, the area A per unit length is equal to $h \times 1 = h$. So (from Das and Seeley, 1975a)

$$F_q = \frac{Q_u}{\gamma A H} = \frac{Q_u}{\gamma h H} = W + K_u \gamma H^2 \tan \phi$$

However $W = \gamma h H$. So

$$F_q = \frac{\gamma h H + K_u \gamma H^2 \tan \phi}{\gamma h H} = 1 + K_u (\frac{H}{h}) \tan \phi \qquad (2\text{-}21)$$

For circular anchors, Equation (2-20) can be modified to the form

$$Q_v = W + \frac{\pi}{2} S_F \gamma h H^2 K_u \tan \phi \qquad (2\text{-}22)$$

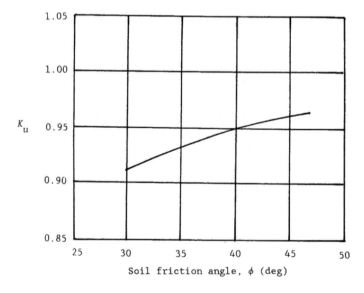

Figure 2-13   Variation of K_u with soil friction angle

where W = the weight of the soil above the circular anchor = $(\frac{\pi}{4} h^2)H\gamma$

h = diameter of the anchor

S_F = shape factor

The shape factor can be expressed as

$$S_F = 1 + m(\frac{H}{h}) \qquad (2-23)$$

where m = coefficient which is a function of the soil friction angle ϕ

Thus, combining Equations (2-22) and (2-23), we obtain

$$Q_u = \frac{\pi}{4}h^2 H\gamma + \frac{\pi}{2}[1 + m(\frac{H}{h})]\gamma hH^2 K_u \tan\phi \qquad (2-24)$$

The breakout factor F_q can be given as (Das and Seeley, 1975a)

$$F_q = \frac{Q_u}{\gamma AH} = \frac{\frac{\pi}{4}h^2 H\gamma + \frac{\pi}{2}[1 + m(\frac{H}{h})]\gamma hH^2 K_u \tan\phi}{\gamma(\frac{\pi}{4} h^2)H}$$

$$= 1 + 2[1 + m(\frac{H}{h})](\frac{H}{h})K_u \tan\phi \qquad (2-25)$$

For *rectangular anchors* having dimensions of $B \times h$, the net ultimate capacity can be expressed as

$$Q_u = W + \gamma H^2(2S_F h + B \quad h)K_u \tan\phi \qquad (2-26)$$

The preceding equation was derived with the assumption that the two end portions of length $h/2$ are governed by the shape factor S_F, while the passive pressure along the central portion of length $B \quad h$ is the same as the strip anchor (Figure 2-14). In Equation (2-26)

$$W = \gamma BhH \qquad (2-27)$$

and

$$S_F = 1 + m(\frac{H}{h}) \qquad (2-23)$$

Thus

$$Q_u = \gamma BhH + \gamma H^2\left\{2[1 + m(\frac{H}{h})]h + B \quad h\right\}K_u \tan\phi \qquad (2-28)$$

The breakout factor F_q can be determined as

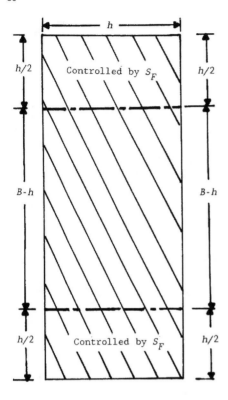

Figure 2-14

$$F_q = \frac{Q_u}{\gamma AH} = \frac{Q_u}{\gamma BhH} \qquad (2\text{-}29)$$

Combining Equations (2-28) and (2-29), we obtain (Das and Seeley, 1975a)

$$F_q = 1 + \left\{ [1 + 2m(\frac{H}{h})](\frac{h}{B}) + 1 \right\}(\frac{H}{h})K_u \tan \phi \qquad (2\text{-}30)$$

The coefficient m given in Equation (2-23) was determined from experimental observations (Meyerhof and Adams, 1968) and its values are given in Table 2-1. In Figure 2-15, m is also plotted as a function of the soil friction angle ϕ.

Experimental observations of Meyerhof and Adams on circular anchors showed that the magnitude of $S_F K_u = [1 + m(H/h)]K_u$ for a given friction angle ϕ increases with H/h to a maximum value at $H/h = (H/h)_{cr}$ and remains constant thereafter as shown in Figure 2-16. This means that, beyond $(H/h)_{cr}$, the anchor behaves as a deep anchor. These $(H/h)_{cr}$ values for square and circular anchors are given in Table 2-2 and also in Figure 2-17.

Table 2-1	Variation of m [Equation (2-23)]

Soil friction angle, ϕ (deg)	m
20	0.05
25	0.1
30	0.15
35	0.25
40	0.35
45	0.5
48	0.6

Table 2-2	Critical Embedment Ratio, $(H/h)_{cr}$, for Square and Circular Anchors

Soil friction angle, ϕ (deg)	$(H/h)_{cr}$
20	2.5
25	3
30	4
35	5
40	7
45	9
48	11

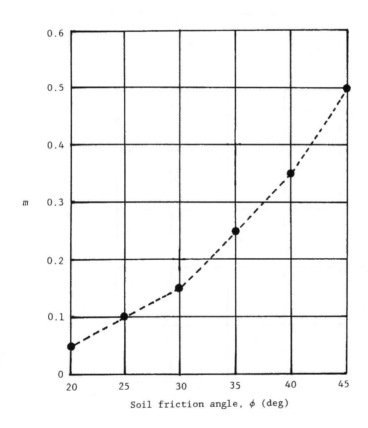

Figure 2-15   Variation of m with soil friction angle ϕ

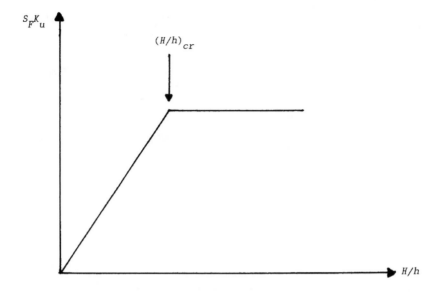

Figure 2-16   Nature of variation of $S_F K_u$ with H/h

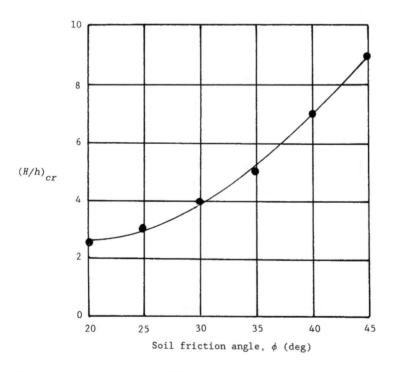

Soil friction angle, ϕ (deg)

Figure 2-17   Variation of $(H/h)_{cr}$ with soil friction angle for square and circular anchors based on the recommendation of Meyerhof and Adams (1968)

Thus for a given value of ϕ, for square ($h = B$) and circular (diameter = h) anchors, we can substitute m (Table 2-1) into Equations (2-25) and (2-30) and calculate the breakout factor (F_q) variation with embedment ratio (H/h). The maximum value of $F_q = F_q^*$ will be attained at $H/h = (H/h)_{cr}$. For $H/h > (H/h)_{cr}$ the breakout factor will remain constant as F_q^*. The variation of F_q with H/h for various values of ϕ made in this manner is shown in Figure 2-18. The variation of the maximum breakout factor F_q^* for deep square and circular anchors with the soil friction angle ϕ is shown in Figure 2-19.

Laboratory experimental observations have shown that the critical embedment ratio (for a given soil friction angle ϕ) increases with the B/h ratio. Meyerhof (1973) has indicated that, for a given value of ϕ,

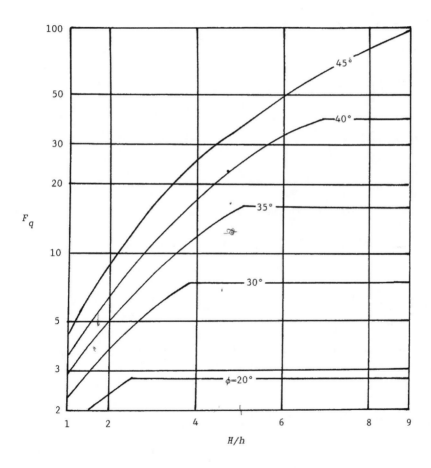

Figure 2-18   Plot of F_q [Equations (2-25) and (2-30)] for square and circular anchors

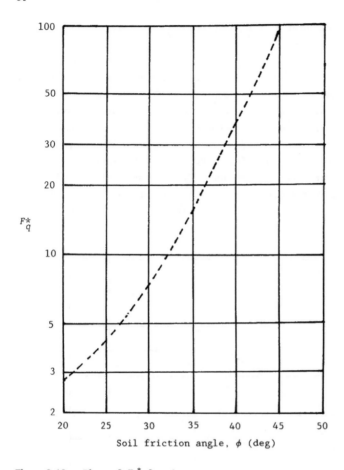

Figure 2-19 Plot of F_q^* for deep square and circular anchors

$$\frac{(\frac{H}{h})_{cr-strip}}{(\frac{H}{h})_{cr-square}} \approx 1.5 \tag{2-31}$$

Based on laboratory model test results, Das and Jones (1982) gave an empirical relationship for the critical embedment ratio of rectangular anchors in the form

$$(\frac{H}{h})_{cr-R} = (\frac{H}{h})_{cr-S}[0.133(\frac{B}{h}) + 0.867] \leq 1.4(\frac{H}{h})_{cr-S} \tag{2-32}$$

where $(H/h)_{cr-R}$ = critical embedment ratio of a rectangular anchor having dimensions of $B \times h$

$(H/h)_{cr-S}$ = critical embedment ratio of a square anchor having dimensions of $h \times h$

Using Equation (2-32) and the $(H/h)_{cr-S}$ values given in Table 2-2, the magnitude of $(H/h)_{cr-R}$ for rectangular anchors can be estimated. These values of $(H/h)_{cr-R}$ can be substituted into Equation (2-30) to determine the variation of $F_q = F_q^*$ with the soil friction angle ϕ. Thus, the uplift capacity of *shallow* and *deep* anchors can be summarized as follows:

For shallow anchors:

$$Q_{u(g)} = F_q \gamma A H + W_a \qquad (2-33)$$

For deep anchors:

$$Q_{u(g)} = F^* \gamma a H + K_o p(H - H_{cr})\bar{\sigma}_o' \tan \phi + W_a \qquad (2-34)$$

where p = perimeter of the anchor shaft
    $H - H_{cr}$ = effective length of the anchor shaft (Figure 2-20)
    $\bar{\sigma}_o'$ = average effective stress between $z = 0$ to $z = H - H_{cr}$
       $= \frac{1}{2}\gamma(H - H_{cr})$
    K_o = at-rest earth pressure coefficient $\approx 1 - \sin \phi$

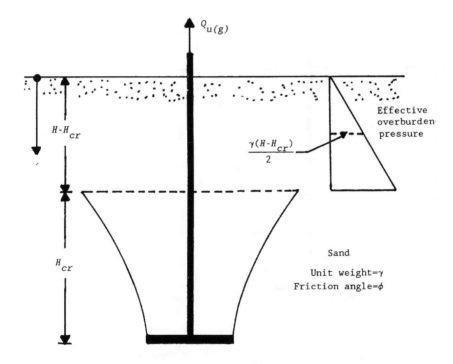

Figure 2-20   Deep horizontal plate anchor

The term $K_o p(H \quad H_{cr})\bar{\sigma}_o'$ tanϕ in Equation (2-34) is the frictional resistance of the shaft. Thus

$$K_o p(H \quad H_{cr})\bar{\sigma}' \tan \phi = \frac{1}{2}\gamma(H \quad H_{cr})^2 p(1 \quad \sin \phi)(\tan \phi) \qquad (2-35)$$

Combining Equations (2-34) and (2-35)

$$Q_{u(g)} = F_q^* \gamma A H + \frac{1}{2}\gamma(H \quad H_{cr})^2 p(1 \quad \sin \phi)(\tan \phi) + W_a \qquad (2-36)$$

2.3.5 Veesaert and Clemence's Theory

Based on laboratory model tests results, Veesaert and Clemence (1977) suggested that for *shallow circular anchors* the failure surface at ultimate load may be approximated as a truncated cone with an apex angle as shown in Figure 2-21. With this type of failure surface, the net ultimate uplift capacity can be given as

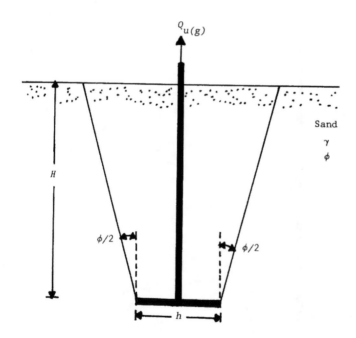

Figure 2-21   Assumption of the failure surface in sand for a circular horizontal plate anchor--Veesaert and Clemence's theory (1977)

$$Q_u = \gamma V + \pi \gamma K_o (\tan \phi)(\cos^2 \frac{\phi}{2}) \left[\frac{hH^2}{2} + \frac{H^3 \tan(\frac{\phi}{2})}{3} \right] \qquad (2\text{-}37)$$

where V = volume of the truncated cone above the anchor

K_o = coefficient of lateral earth pressure

$$V = \frac{\pi H}{3}[h^2 + (h + 2H \tan \frac{\phi}{2})^2 + (h)(h + 2H \tan \frac{\phi}{2})]$$

$$= \frac{\pi H}{3}[3h^2 + 4H^2 \tan^2(\frac{\phi}{2}) + 6Hh \tan(\frac{\phi}{2})] \qquad (2\text{-}38)$$

Substituting Equation (2-38) into Equation (2-37), we obtain

$$Q_u = \frac{\pi \gamma H}{3}[3h^2 + 4H^2 \tan^2(\frac{\phi}{2}) + 6Hh \tan(\frac{\phi}{2})]$$

$$+ \pi \gamma K_o (\tan \phi)(\cos^2 \frac{\phi}{2}) \left[\frac{hH^2}{2} + \frac{H^3 \tan(\frac{\phi}{2})}{3} \right] \qquad (2\text{-}39)$$

The breakout factor can now be determined as

$$F_q = \frac{Q_u}{\gamma AH} = \frac{Q_u}{\gamma(\frac{\pi}{4} h^2)H} \qquad (2\text{-}40)$$

Combining Equations (2-39) and (2-40)

$$F_q = \left\{ 4K_o (\tan \phi)(\cos^2 \frac{\phi}{2})(\frac{H}{h})^2 \left[\frac{0.5}{(\frac{H}{h})} + \frac{\tan(\frac{\phi}{2})}{3} \right] \right\}$$

$$+ [4 + 8(\frac{H}{h})\tan(\frac{\phi}{2}) + 5.333(\frac{H}{h})^2 \tan^2(\frac{\phi}{2})] \qquad (2\text{-}41)$$

Veesaert and Clemence (1977) suggested that the magnitude of K_o may vary between 0.6 to 1.5 with an average value of about 1. Figure 2-22 shows the plot of F_q versus H/h with $K_o = 1$. In this plot it is assumed that $(H/h)_{cr}$ is the same as that proposed by Meyerhof and Adams (1968) and given in Table 2-2. For $H/h \le (H/h)_{cr}$ the magnitude of $F_q = F_q^* = $ constant. A comparison of the plots shown in Figures 2-18 and 2-22 shows the following:

1. For ϕ up to about 35° with $K_o = 1$, Equation (2-41) yields higher values of F_q compared to that calculated by using Equation (2-30).

2. For $\phi = 40°$ and similar H/h ratios, Equations (2-30) and (2-41) yield practically the same values of F_q.

3. For $\phi > 40°$ the values of F_q calculated by using Equation (2-41) are smaller than those calculated by using Equation (2-30).

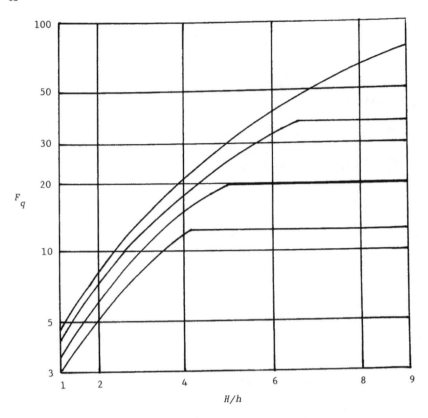

Figure 2-22   Variation of F_q for shallow circular anchors [Equation (2-41)]

2.3.6   Vesic's Theory

Vesic (1965) studied the problem of an explosive point charge expanding a spherical cavity close to the surface of a semi-infinite, homogeneous, isotropic solid (in this case, the soil).   Now, referring to Figure 2-23, it can be seen that, if the distance H is small enough, there will be an ultimate pressure p_o that will shear away the soil located above the cavity   At that time the diameter of the spherical cavity is equal to h.   The slip surfaces ab and cd will be tangent to the spherical cavity at a and c.   At points b and d, they make an angle $\alpha = 45$ $\phi/2$.   Now, for equilibrium, summing the components of forces in the vertical direction we can determine the ultimate pressure p_o in the cavity.   Forces that will be involved are:

1. Vertical component of the force inside the cavity, P_V
2. Effective self-weight of the soil, $W = W_1 + W_2$
3. Vertical component of the resultant of internal forces, F_V

For a c-ϕ soil, we can thus determine that

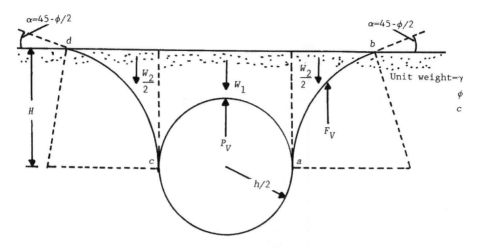

Figure 2-23 Vesic's theory of expansion of cavities

$$P_o = c\overline{F}_c + \gamma H \overline{F}_q \qquad (2\text{-}42)$$

where $\overline{F}_q = 1.0 \quad \dfrac{2}{3}\left[\dfrac{(\frac{h}{2})}{H}\right] + A_1\left[\dfrac{H}{(\frac{h}{2})}\right] + A_2\left[\dfrac{H}{(\frac{h}{2})}\right]^2 \qquad (2\text{-}43)$

$$\overline{F}_c = A_3\left[\dfrac{H}{(\frac{h}{2})}\right] + A_4\left[\dfrac{H}{(\frac{h}{2})}\right] \qquad (2\text{-}44)$$

where A_1, A_2, A_3, A_4 = functions of the soil friction angle ϕ

For granular soils $c = 0$. So

$$P_o \quad \gamma H \overline{F}_q \qquad (2\text{-}45)$$

Vesic (1971) applied the preceding concept to determine the ultimate uplift capacity of shallow circular anchors. In Figure 2-23, consider that the circular anchor plate *ab*, having a diameter h, is located at a depth H below the ground surface. If the hemispherical cavity above the anchor plate is filled with soil it will have a weight of (Figure 2-24)

$$W_3 = \frac{2}{3}\pi\left(\frac{h}{2}\right)^3\gamma \qquad (2\text{-}46)$$

This weight of soil will increase the pressure by p_1, or

$$p_1 = \frac{W_3}{\pi\left(\frac{h}{2}\right)^2} = \frac{\left(\frac{2}{3}\right)\pi\left(\frac{h}{2}\right)^3\gamma}{\pi\left(\frac{h}{2}\right)^2} = \frac{2}{3}\gamma\left(\frac{h}{2}\right)$$

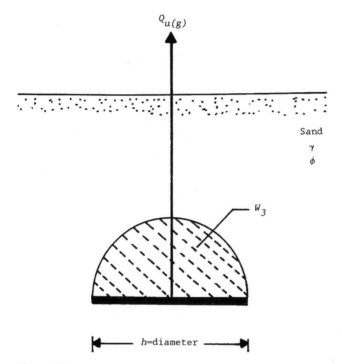

Figure 2-24

If the anchor is embedded in a cohesionless soil ($c = 0$), then the pressure p_1 should be added to Equation (2-43) to obtain the force per unit area of the anchor, q_u, needed for complete pullout. Thus

$$q_u = \frac{Q_u}{A} = \frac{Q_u}{\frac{\pi}{2}(h)^2} = p_o + p_1 = \gamma H(\overline{F}_q) + \frac{2}{3}\gamma(\frac{h}{2}) = \gamma H\left[\overline{F}_q + \frac{(\frac{2}{3})(\frac{h}{2})}{H}\right] \qquad (2\text{-}47)$$

or

$$q_u = \frac{Q_u}{A} = \gamma H\left\{1 + A_1\left[\frac{H}{(\frac{h}{2})}\right] + A_2\left[\frac{H}{(\frac{h}{2})}\right]^2\right\} = \gamma H F_q \qquad (2\text{-}48)$$

$$\uparrow$$

Breakout factor

The variation of the breakout factor F_q for *shallow circular anchor plates* is given in Table 2-3 (also Figure 2-25). In a similar manner, using the analogy of the expansion of long cylindrical cavities, Vesic determined the variation of the breakout factor F_q for *shallow strip anchors*. These values are given in Table 2-4 and are also plotted in Figure 2-26.

Table 2-3 Vesic's (1971) Breakout Factor, F_q, for Circular Anchors

Soil friction angle, ϕ (deg)	H/h				
	0.5	1.0	1.5	2.5	5.0
0	1.0	1.0	1.0	1.0	1.0
10	1.18	1.37	1.59	2.08	3.67
20	1.36	1.75	2.20	3.25	6.71
30	1.52	2.11	2.79	4.41	9.89
40	1.65	2.41	3.30	5.45	13.0
50	1.73	2.61	3.56	6.27	15.7

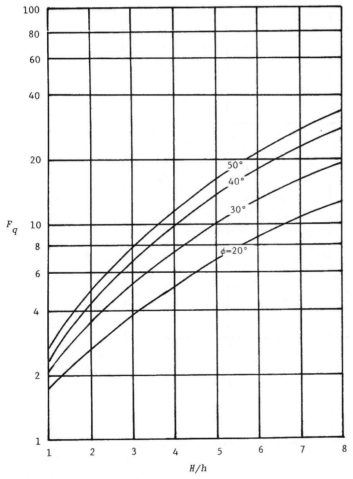

Figure 2-25 Vesic's (1971) breakout factor F_q for shallow circular anchors

Table 2-4   Vesic's (1971) Breakout Factor, F_q, for Strip Anchors

Soil friction angle, ϕ (deg)	H/h				
	0.5	1.0	1.5	2.5	5.0
0	1.0	1.0	1.0	1.0	1.0
10	1.09	1.16	1.25	1.42	1.83
20	1.17	1.33	1.49	1.83	2.65
30	1.24	1.47	1.71	2.19	3.38
40	1.30	1.58	1.87	2.46	3.91
50	1.32	1.64	2.04	2.6	4.2

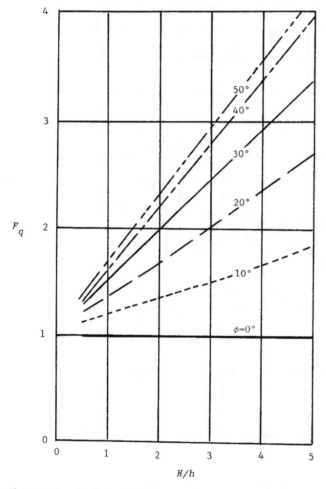

Figure 2-26   Vesic's (1971) breakout factor for shallow strip anchors

2.3.7 Saeedy's Theory

An ultimate holding capacity theory for *circular plate anchors* embedded in sand was proposed by Saaedy (1987) in which the trace of the failure surface was assumed to be an arc of a logarithmic spiral as shown in Figure 2-27. According to this solution, for shallow anchors the failure surface extends to the ground surface. However, for deep anchors (that is, $H > H_{cr}$), the failure surface extends to a distance of H_{cr} above the anchor plate. Based on this analysis, Saeedy (1987) proposed the net ultimate uplift capacity in a non-dimensional form ($Q_u/\gamma Hh^2$) for various values of ϕ and the H/h ratio. The author has converted the solution into a plot of breakout factor $F_q = Q_u/\gamma AH$ (A = area of the anchor plate) versus the soil friction angle ϕ as shown in Figure 2-28. According to Saeedy (1987), during the anchor pullout the soil located above the anchor gradually becomes compacted, in turn increasing the shear strength of the soil and, hence, the net ultimate uplift capacity. For that reason, he introduced an empirical *compaction factor* which is given in the form

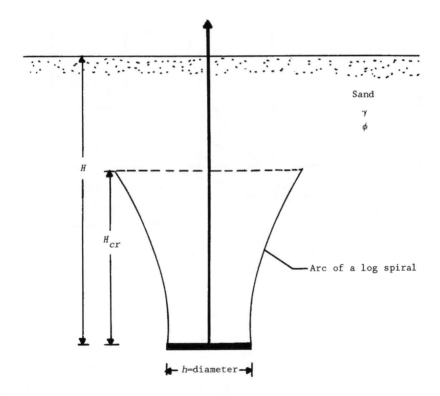

Figure 2-27   Saeedy's theory for circular plate anchors

44

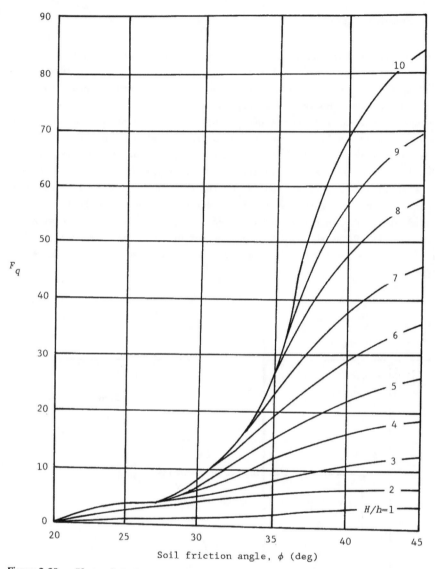

Figure 2-28   Plot of F_q based on Saeedy's theory

$$\mu = 1.044 D_r + 0.44 \qquad\qquad (2\text{-}49)$$

where μ = compaction factor

D_r = relative density of compaction

Thus, the actual net ultimate capacity can be expressed as

$$Q_{u(actual)} = (F_q \gamma A H)\mu \qquad\qquad (2\text{-}50)$$

2.3.8 Discussion of Various Theories

Based on various theories presented in this section (Section 2.3), we can make some general observations:

1. All existing theories, except that of Meyerhof and Adams (1968), are for axisymmetric case (that is, for use in the case of *circular* anchors). Meyerhof and Adams' theory addresses the case of *rectangular* anchors.

2. Most theories assume that the shallow anchor condition exists for $H/B \leq 5$. Meyerhof and Adams' theory provides a critical embedment ratio $(H/h)_{cr}$ for *square and circular* anchors as a function of the soil friction angle.

3. Experimental observations generally tend to show that, for shallow anchors embedded in loose sand, Balla's theory (1961) overestimates the net ultimate uplift capacity. However better agreement is obtained for anchors embedded in dense soil.

4. Vesic's theory (1971) is generally fairly accurate in estimating the net ultimate uplift capacity for shallow anchors in loose sand. However, laboratory experimental observations have shown that for shallow anchors embedded in dense sand, this theory can underestimate the actual capacity by as much as 100% or more.

5. Mariupol'skii's theory (1965) suggests that, for calculation of the net ultimate capacity, the lower of the two values obtained from Equations (2-14) and (2-15) should be used. The reason for such recommendation is due to the fact that the critical embedment was not clearly established in the theory.

Figure 2-29 shows a comparison of some published laboratory experimental results for the net ultimate uplift capacity of *circular anchors* with the theories of Balla, Vesic, and Meyerhof and Adams. Table 2-5 gives the references to the laboratory experimental curves shown in Figure 2-29. In developing the theoretical plots for $\phi = 30°$ (loose sand condition) and $\phi = 45°$ (dense sand condition) the following procedures have been used.

1. According to Balla's theory (1961), from Equation (2-9) for circular anchors

$$Q_u = H^3 \gamma \underbrace{[F_1 + F_3]}_{\text{Figure 2-7}}$$

So

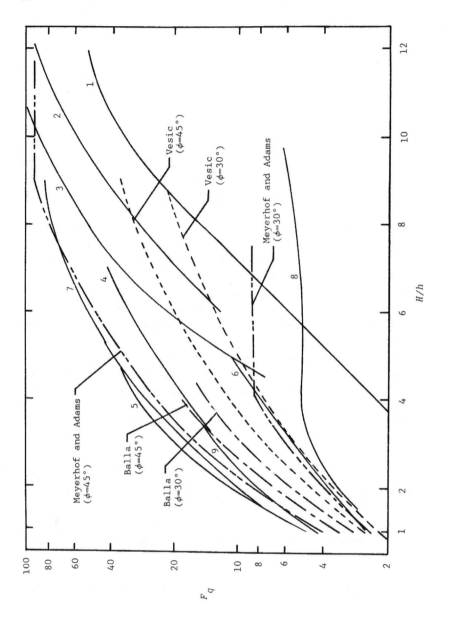

Figure 2-29 Comparison of theories with laboratory experimental results for circular anchor plates

Table 2-5   References to Laboratory Experimental Curves Shown in Figure 2-29

Curve No.	Reference	Circular anchor diameter, h (mm)	Soil properties
1	Baker and Kondner (1966)	25.4	$\phi = 42°$; $\gamma = 17.61$ kN/m^3
2	Baker and Kondner (1966)	38.1	$\phi = 42°$; $\gamma = 17.61$ kN/m^3
3	Baker and Kondner (1966)	50.8	$\phi = 42°$; $\gamma = 17.61$ kN/m^3
4	Baker and Kondner (1966)	76.2	$\phi = 42°$; $\gamma = 17.61$ kN/m^3
5	Sutherland (1965)	38.1 to 152.4	$\phi = 45°$
6	Sutherland (1965)	38.1 to 152.4	$\phi = 31°$
7	Esquivel-Diaz (1967)	76.2	$\phi \approx 43°$; $\gamma = 14.81$ kN/m^3 to 15.14 kN/m^3
8	Esquivel-Diaz (1967)	76.2	$\phi = 33°$ $\gamma = 12.73$ kN/m^3 to 12.89 kN/m^3
9	Balla (1961)	61 to 119.4	Dense sand

$$F_1 + F_3 = \frac{Q_u}{\gamma H^3} = \frac{(\frac{\pi}{4} h^2)Q_u}{\gamma H^3 (\frac{\pi}{4} h^2)} = \frac{[\frac{\pi}{4}(\frac{h}{H})^2]Q_u}{\gamma HA}$$

or

$$F_q = \frac{Q_u}{\gamma AH} = \frac{F_1 + F_3}{(\frac{\pi}{4})(\frac{h}{H})^2} \qquad (2\text{-}51)$$

So, for a given soil friction angle, the sum of $F_1 + F_3$ has been obtained from Figure 2-7 and the breakout factor has been calculated for various values of H/h and have been plotted in Figure 2-29.

2.   For Vesic's theory (1971), the variations of F_q versus H/h for circular anchors have been given in Table 2-3. These values of F_q have also been plotted in Figure 2-29.

3.   The breakout factor relationship for circular anchors based on Meyerhof and Adams' theory (1968) is given in Equation (2-25). Using $K_u \approx 0.95$, the variations of F_q with H/h have been calculated and are plotted in Figure 2-29.

Based on the comparison between the theories and the laboratory experimental results shown in Figure 2-29, it appears that Meyerhof and Adams' theory (1968) is more applicable to a wide range of anchors and provides as good an estimate as any for the net ultimate uplift capacity. So this theory is recommended for use. However it needs to be kept in mind that the majority of the experimental results presently available in the literature for comparison with the theory are from laboratory model tests. When applying these results to the design of an actual foundation, the *scale effect* needs to be taken into consideration. For that reason, a judicious choice is necessary in selecting the value of the soil friction angle ϕ.

Example

2.1

Consider a circular anchor plate embedded in sand. Given, for the anchor: diameter, $h = 0.3$ m; depth of embedment, $H = 1.2$ m. Given, for the sand: unit weight, $\gamma = 17.4$ kN/m^3; friction angle, $\phi = 35°$ Using Balla's theory, calculate the net ultimate uplift capacity.

Solution

From Equation (2-9)

$$Q_u = H^3 \gamma [F_1 + F_3]$$

From Figure 2-7, for $\phi = 35°$ and $H/h = 1.2/0.3 = 4$, the magnitude of $F_1 + F_3 \approx 0.725$. So

$$Q_U = (1.2)^3 (17.4)(0.725) = \textbf{21.8 kN}$$

Example

2.2

Redo Example Problem 2.1 using Vesic's theory.

Solution

From Equation (2-48)

$$Q_u = A\gamma H F_q$$

From Figure 2-25, for $\phi = 35°$ and $H/h = 4$, F_q is about 9. So

$$Q_u = [(\tfrac{\pi}{4})(0.3)^2](17.4)(1.2)(9) = \textbf{13.28 kN}$$

Example

2.3

Redo Example Problem 2.1 using Meyerhof and Adams' theory.

Solution

From Equation (2-25)

$$F_q = 1 + 2[1 + m(\frac{H}{h})](\frac{H}{h})K_u \tan \phi$$

For $\phi = 35°$, $m = 0.25$ (Table 2-2). So

$$F_q = 1 + 2[1 + (0.25)(4)](4)(0.95)(\tan 35) = 11.64$$

So

$$Q_u = F_q \gamma AH = (11.64)(17.4)[(\frac{\pi}{4})(0.3)^2](1.2) = \textbf{17.18 kN}$$

Example

2.4

Redo Example Problem 2.1 using Veesaert and Clemence's theory. Use $K_o = 1$.

Solution

From Equation (2-41)

$$F_q = \left\{ 4K_o(\tan \phi)(\cos^2 \frac{\phi}{2})(\frac{H}{h})^2 \left[\frac{0.5}{(\frac{H}{h})} + \frac{\tan(\frac{\phi}{2})}{3} \right] \right\} + [4 + 8(\frac{H}{h})(\tan \frac{\phi}{2})$$

$$+ 5.333(\frac{H}{h})^2(\tan^2 \frac{\phi}{2})]$$

Given: $\phi = 35°$, $H/h = 4$; $K_o = 1$. So

$$F_q = \left\{ [(4)(1)(\tan 35)(\cos^2 17.5)(4)^2 [\frac{0.5}{4} + \frac{\tan(17.5)}{3}] \right\}$$

$$+ [4 + (8)(4)(\tan 17.5) + 5.333(4)^2(\tan^2 17.5)] = 32.4$$

So

$$Q_u = F_q \gamma AH = (32.4)(17.4)[(\frac{\pi}{4})(0.3)^2](1.2) = \textbf{47.82 kN}$$

2.4   LOAD-DISPLACEMENT RELATIONSHIP

In order to determine the allowable net ultimate uplift capacity of plate anchors, two different procedures can be adopted:

1.   Use of a *tentative factor of safety*, F_s, based on the uncertainties of determination of the soil shear strength parameters and other associated factors.   For this type of analysis

$$Q_{u(all)} = \frac{Q_u}{F_s} \qquad (2\text{-}52)$$

2   Use of a *load-displacement relationship*.   In this method, the allowable net ultimate uplift capacity is calculated which corresponds to a predetermined allowable vertical displacement of the anchor.

Das and Puri (1989) investigated the load-displacement relationship of *shallow horizontal square and rectangular plate anchors* embedded in medium and dense sand.   For these laboratory model tests, the width of the anchor plate (h) was kept at 50.8 mm.   The length-to-width ratios of the anchors (B/h) were varied from one to three, and the H/h ratios were varied from one to five. Based on their laboratory observations, the net load Q versus vertical displacement Δ plots can be of two types, as shown in Figure 2-30.   In Type I, the net load increases with displacement up to a maximum value at which sudden pullout occurs.   The maximum load in this case is the net ultimate uplift capacity Q_u In Type II, the net load increases with the vertical displacement fairly rapidly up to a certain point, beyond which the load-displacement relationship becomes practically linear.   For this case, the net ultimate uplift capacity is defined as the point where the slope of the Q versus Δ plot becomes minimum.   The vertical displacement which corresponds to load Q_u, is defined as Δ_u in Figure 2-30.

Figure 2-31 shows the magnitudes of Δ_u for anchors with various B/h ratios placed at varying embedment ratios (H/h).   It needs to be pointed out that, for tests conducted in medium sand, the relative density of compaction D_r was about 48%.   Similarly, for tests conducted in dense sand, the average value of D_r was about 73%.   With their experimental results, Das and Puri (1989) proposed a nondimensional empirical load-displacement relationship for shallow plate anchors which is of the form

$$\overline{Q} = \frac{\overline{\Delta}}{a + b\overline{\Delta}} \qquad (2\text{-}53)$$

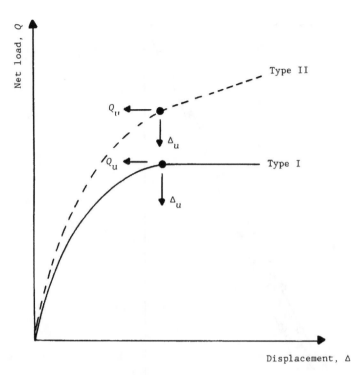

Figure 2-30 Nature of load versus displacement plots

where $\bar{Q} = \dfrac{Q}{Q_u}$ $\hspace{4cm}$ (2-54)

$\bar{\Delta} = \dfrac{\Delta}{\Delta_u}$ $\hspace{4cm}$ (2-55)

$\quad \Delta$ = anchor displacement at net uplifting load Q

$\quad a, b$ = constants

The constants a and b are approximately equal to 0.175 and 0.825, respectively, and they are not functions of the relative density of compaction. From Equation (2-53), it follows that

$$\frac{\bar{\Delta}}{\bar{Q}} = a + b\bar{\Delta}$$ $\hspace{3cm}$ (2-56)

The preceding equation implies that a plot of $\bar{\Delta}/\bar{Q}$ versus $\bar{\Delta}$ will be approximately linear. Figures 2-32 and 2-33 show the plot of $\bar{\Delta}/\bar{Q}$ versus $\bar{\Delta}$ for laboratory model tests conducted by Das and Puri (1989) in medium and dense sand.

52

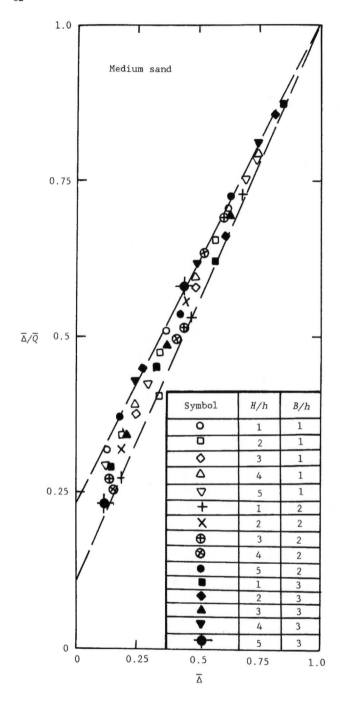

Figure 2-32   Variation of $\overline{\Delta}/\overline{Q}$ with $\overline{\Delta}$ for medium sand based on the
model tests of Das and Puri (1989)

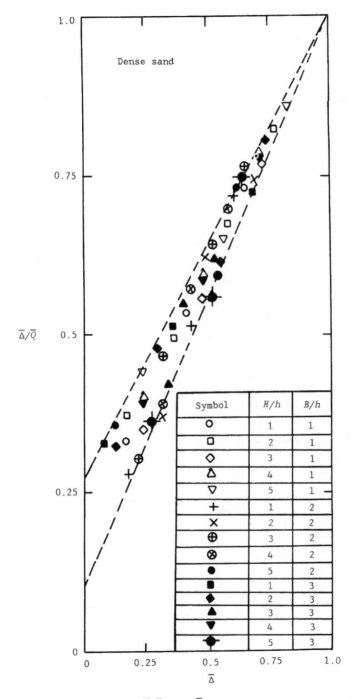

Figure 2-33   Variation of $\overline{\Delta}/\overline{Q}$ with $\overline{\Delta}$ for dense sand based on the model tests of Das and Puri (1989)

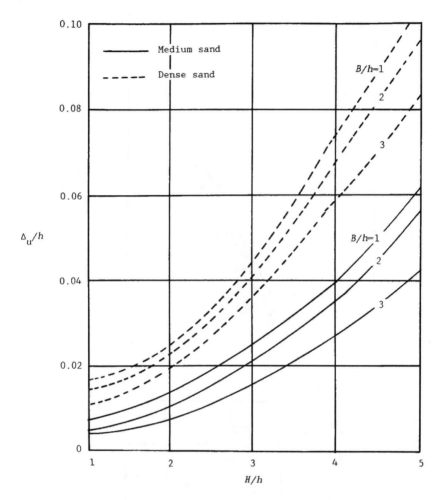

Figure 2-31 Variation of Δ_u/h versus H/h based on the model tests of Das and Puri (1989) (*Note*: $h = 50.8$ m)

Example

2.5

Consider a shallow rectangular anchor embedded in sand, for which the following is given: $h = 0.3$ m, $B = 0.9$ m, $H = 1.2$ m. For the sand, given: $\gamma = 18$ kN/m^3, $\phi = 35°$ Estimate:

a. the net ultimate uplift capacity using the theory of Meyerhof and Adams,

b. the anchor displacement at ultimate load, and

c. the net load Q at an anchor displacement of $0.5\Delta_u$

Solution

Part a.   For this case

$$\frac{B}{h} = \frac{0.9}{0.3} = 3; \frac{H}{h} = \frac{1.2}{0.3} = 4$$

From Table 2-2, $H/h < (H/h)_{cr}$ for $\phi = 35°$ So it is a shallow anchor. From Equations (2-29) and (2-30)

$$Q_u = F_q \gamma BhH$$

$$F_q = 1 + \left\{[1 + 2m(\frac{H}{h})](\frac{h}{B}) + 1\right\}(\frac{H}{h})K_u \tan \phi$$

For $\phi = 35°$, the value of m is 0.25.   Assuming $K_u \approx 0.95$, we can calculate F_q So

$$F_q = 1 + \left\{[1 + (2)(0.25)(4)](\frac{1}{3}) + 1\right\}(4)(0.95)(\tan 35) = 6.32$$

So

$$Q_u = F_q \gamma BhH = (6.32)(18)(0.9)(0.3)(1.2) = \textbf{36.86 kN}$$

Part b.   Consider the sand as loose.   From Figure 2-31, for $B/h = 3$ and $H/h = 4$, the value of $\Delta_u/h \approx 0.06$.   So

$$\Delta_u \approx (0.06)(0.3) = 0.18 \text{ m} = \textbf{180 mm}$$

Part c.   From Equation (2-53)

$$\bar{Q} = \frac{\bar{\Delta}}{a + b\bar{\Delta}}; \bar{\Delta} = \frac{\Delta}{\Delta_u} = 0.5$$

So

$$\bar{Q} = \frac{0.5}{0.175 + (0.825)(0.5)} = 0.851; \bar{Q} = \frac{Q}{Q_u} = 0.851$$

So

$$Q = (0.851)(36.86) = \textbf{31.37 kN}$$

2.5   ANCHORS SUBJECTED TO REPEATED LOADING

Horizontal anchors are sometimes used to moor surface vessels or buoys as well as semi-submersible or submersible structures.   These anchors may be subjected to a combination of sustained and repeated loads. The application of repeated loads may create a progressive accumulative cyclic strain which will ultimately lead to the uplift of the anchor. Very few studies are available at the present time to evaluate the effect of repeated loads on anchors. Andreadis, Harvey and Burley (1978) studied the behavior of model circular anchor plates embedded in dense saturated sand and subjected to cyclic loading. For this study the embedment ratio H/h was kept at 12 (that is, deep anchor condition). The cyclic load was sinusoidal in nature with 10-second duration cycles (Figure 2-34a). In some tests, the cyclic load Q_c was applied alone as shown in Figure 2-34b. Also some tests were conducted with an initial application of a sustained static load Q_s and then a cyclic load of magnitude Q_c, and the results of these tests are shown in Figure 2-35. In Figure 2-35, the relative anchor movement is defined as

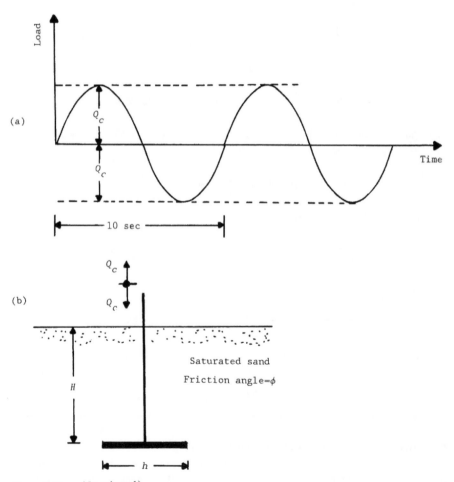

Figure 2-34   (Continued)

57

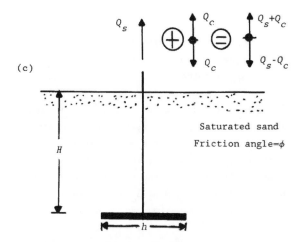

Figure 2-34   Details of the model tests of Andreadis et al. (1978)
on deep circular anchor plates

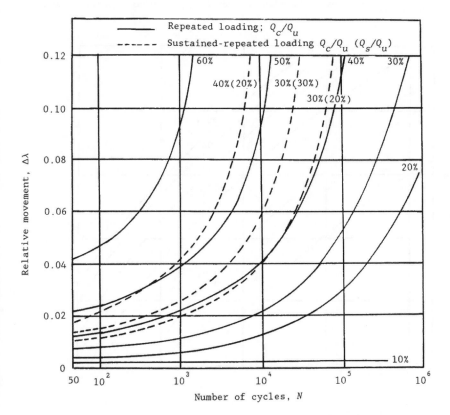

Figure 2-35   Relative anchor movement versus number of cycles--
dense sand; $H/h = 12$; circular anchor (after
Andreadis et al., 1978)

58

$$\Delta\lambda = \frac{\text{Uplift of anchor, } \Delta}{\text{Anchor diameter, } h} \tag{2-57}$$

It can be seen from Figure 2-35 that, for a given magnitude of Q_c/Q_u, the relative anchor displacement $\Delta\lambda$ increased with the number of cycles.

Based on their model tests, Andreadis, Harvey and Burley (1978) suggested that, when the cyclic relative anchor displacement is kept below about half the relative movement to failure in static pullout tests, there is essentially no reduction in strength due to cyclic loading. For that reason, a plot of Q_c/Q_u versus number of cyclic load applications for various values of $\Delta\lambda$ is shown in Figure 2-36, essentially obtained from the experimental results shown in Figure 2-35. So, if the ultimate displacement Δ_u at ultimate static load Q_u is known, one can calculate the allowable maximum value $\Delta\lambda$ as

$$\Delta\lambda_{(allowable)} \approx \frac{1}{2}\Delta_u \tag{2-58}$$

Once $\Delta\lambda_{(allowable)}$ is known, the magnitude of Q_c/Q_u and thus Q_c, corresponding

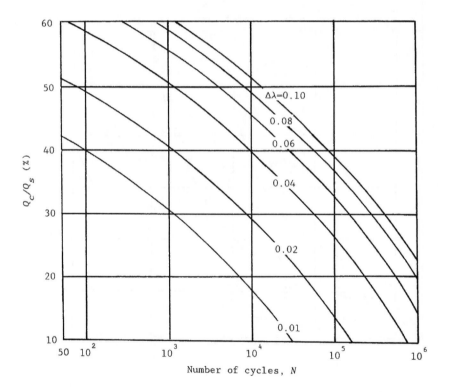

Figure 2-36   Relative cyclic load versus number of cycles--dense sand; $H/h = 12$; circular anchor (after Andreadis et al., 1978)

to the number of load application cycles during the life span of the anchor, can be estimated.

2.6 UPLIFT CAPACITY OF SHALLOW GROUP ANCHORS

When anchors placed in a group are subjected to an uplifting load, the net ultimate uplift capacity of the group may possibly be smaller than the net ultimate uplift capacity of a single anchor times the number of anchors in the group. This condition arises when the center-to-center spacing of the anchor is small and when, during the anchor uplift, there is interference of the failure zones in soil. Figure 2-37 shows a group of anchors located at a shallow depth H. All of the anchors are circular in shape, and the center-to-center spacing of the anchors is equal to s. In the plan of the anchor group there are m number of rows and n columns. The gross ultimate uplift capacity of the anchor group, $Q_{ug(g)}$, can be given as

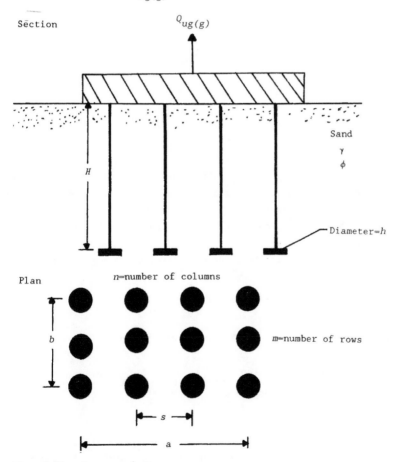

Figure 2-37   Group anchors

$$Q_{ug(g)} = Q_{ug} + W_g \qquad (2\text{-}59)$$

where Q_{ug} = net ultimate uplift capacity of the group

W_g = effective self-weight of anchors and the shafts

Meyerhof and Adams (1968) derived a theoretical relationship for the net ultimate capacity of group anchors, according to which

$$Q_{ug} = \gamma H^2 [a + b + S_F(\tfrac{\pi}{2})h]K_u \tan \phi + W_s \qquad (2\text{-}60)$$

where S_F = shape factor

K_u = nominal uplift coefficient

W_s = effective weight of the sand located above the anchor group

$a = s(n - 1)$ (see Figure 2-37)

$b = s(m - 1)$ (see Figure 2-37)

The shape factor S_F is given by the same relationship as in Equation (2-23), or

$$S_F = 1 + m(\tfrac{H}{h})$$
$$\uparrow$$

See Table 2-1

The nominal uplift coefficient K_u is the same as shown in Figure 2-13 and may be taken as approximately 0.95 for all values of the soil friction angle ϕ. In deriving Equation (2-60), it is assumed that the passive pressure along the curved portion of the perimeter of the group is governed by the shape factor S_F, and the passive earth pressure along the straight portions is the same as for strip anchors.

In the conventional manner, the group efficiency η can now be defined as

$$\eta = \frac{Q_{ug}}{mnQ_u} \qquad (2\text{-}61)$$

Thus, combining Equations (2-60), (2-61) and (2-22), we obtain

$$\eta(\%) = \left\{ \frac{\gamma H^2 [a + b + S_F(\tfrac{\pi}{2})h]K_u \tan \phi + W_s}{mn[(\tfrac{\pi}{2})S_F \gamma h H^2 K_u \tan \phi + W]} \right\} (100) \leq 100\% \qquad (2\text{-}62)$$

In order to investigate the applicability of the preceding equation, Das and Jin Kaun (1987) conducted a limited number of laboratory model tests in compacted sand at a relative density of 68% with an angle of friction of 37°.

Figures 2-38 and 2-39 show the model test results for group efficiency for the cases of $H/h = 4$ and 6, respectively. The theoretical variations of the group efficiency with the center-to-center spacing of anchors are also shown in Figures 2-38 and 2-39. A comparison of the theoretical and experimental results show that, for a given anchor configuration and H/h, the s/h ratio at which $\eta = 100\%$ is approximately twice that predicted by the theory. However the general trend of the actual variation of η versus s/h for a given anchor configuration is similar to that predicted by the theory.

2.7   SPREAD FOUNDATIONS UNDER UPLIFT

Spread foundations constructed for electric transmission towers are subjected to uplifting force. The uplift capacity of such foundations can be estimated by using the same relationship described in this chapter. During the construction of such foundations, the embedment ratio H/h is usually kept at 3 or less. The native soil is first excavated for foundation construction. Once the foundation construction is finished, the excavation is backfilled and compacted. The degree of compaction of the backfill material plays an important role in the actual net ultimate uplift capacity of the foundation. Kulhawy, Trautmann and Nicolaides (1987) conducted several laboratory model

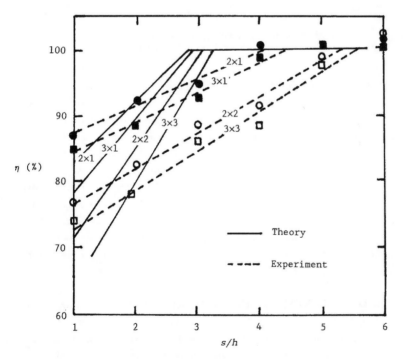

Figure 2-38   Variation of η versus s/h for group piles--relative density = 68%; $H/h = 4$ (after Das and Jin-Kaun, 1987)

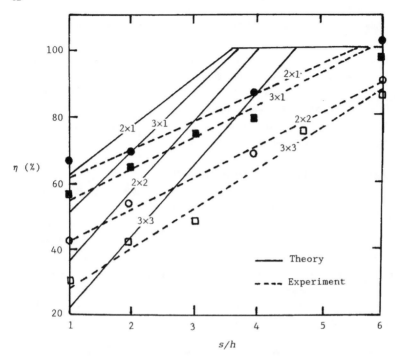

Figure 2-39   Variation of η versus s/h for group piles--relative density = 68%; H/h = 6 (after Das and Jin-Kaun, 1987)

tests to observe the effect of the degree of compaction of the backfill compared to the native soil. According to their observations, in most cases, at ultimate load failure in soil take place by side shear as shown in Figure 2-40. However, wedge or combined shear failure occurs for foundations with H/h < about 2 in medium to dense native soil where the backfill is at least 85% as dense as the native soil (Figure 2-41). Figure 2-42 shows the effect of backfill compaction on the breakout factor F_q when the native soil is loose. Similarly, Figure 2-43 is for the case where the native soil is dense. Based on the observations of Kulhawy et al. (1987), this study shows that the compaction of the backfill has a great influence on the breakout factor of the foundation, and the net ultimate uplift capacity is greatly increased with the degree of backfill compaction.

2.8   INCLINED LOAD RESISTANCE OF HORIZONTAL PLATE ANCHORS

Das and Seeley (1975b) conducted a limited number of model tests to observe the nature of variation of the ultimate uplifting load of *horizontal square plate anchors* embedded in loose sand and subjected to inclined pull. The plate anchor used for the tests was 61 mm × 61 mm. The friction angle of the sand for the density of compaction at which tests were conducted was 31° For

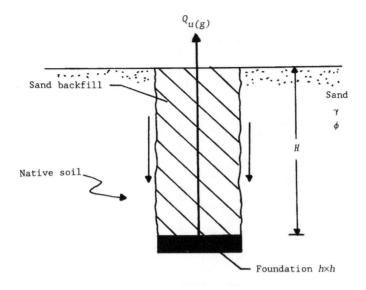

Figure 2-40 Failure by side shear

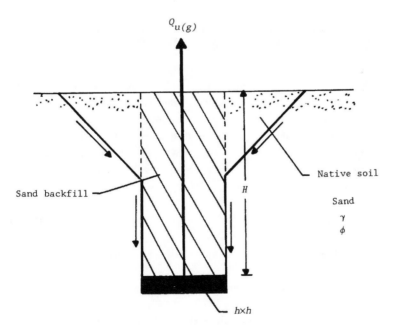

Figure 2-41 Wedge or combined shear failure

64

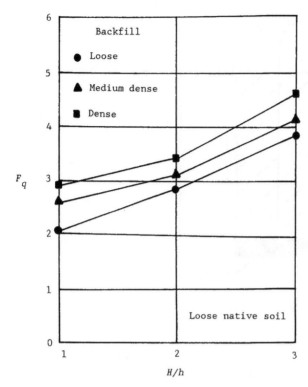

Figure 2-42   Effect of backfill on breakout factor (*Note*: square
            foundation; loose native soil) (after Kulhawy et al.,
            1987)

this study the pullout load on the anchor was applied by a cable that can allow full rotation of the anchor during pullout. Such conditions may arise to
moor surface vessels or buoys, and also semi-submersible or submerged structures. Figure 2-44 shows an anchor plate embedded at a depth H and subjected
to a gross ultimate uplift load $Q_{u-\psi(g)}$, with the load inclined at an angle ψ
with respect to the vertical. The net ultimate uplift capacity can thus be
given as

$$Q_{u-\psi} = Q_{u-\psi(g)} \quad W_a \cos \psi \qquad (2-63)$$

where $Q_{u-\psi}$ = net ultimate uplift capacity measured in the direction of the
            load application

W_a = effective weight of the anchor

Figure 2-45 shows the variation of $Q_{u-\psi}$ with the angle of load inclination ψ
for H/h = 1, 2, 2.5 and 4.5. From this figure it can be seen that, for $\psi \leq$
45°, the magnitude of $Q_{u-\psi}$ increases with the increase of the load

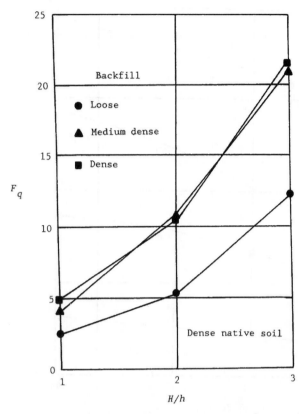

Figure 2-43 Effect of backfill on breakout factor (*Note*: square foundation; dense native soil) (after Kulhawy et al., 1987)

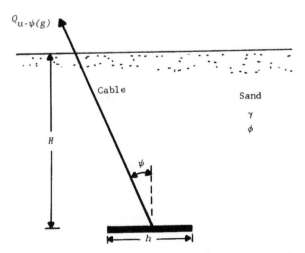

Figure 2-44 Inclined uplifting load on horizontal plate anchor

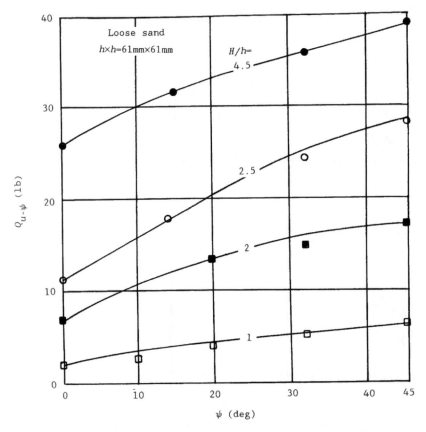

Figure 2-45   Effect of load inclination on $Q_{u-\psi}$ (after Das and Seeley, 1975b)

inclination.   Also, as the embedment ratio H/h increases, the ratio $Q_{u-\psi}/Q_{u-\psi=0}$ decreases (for a given value of ψ).

ANCHORS IN CLAY (ϕ = 0 CONDITION)

2.9   ULTIMATE UPLIFT CAPACITY

Theoretical and experimental research results presently available for determination of the net ultimate uplift capacity of plate anchor embedded in saturated clay soil are rather limited. In this section the results of the existing studies will be reviewed.

Figure 2-46 shows a plate anchor embedded in a saturated clay at a depth H below the ground surface. The width of the anchor plate is equal to h and the undrained cohesion of the clay is c_u In soft saturated clay, when the anchor is subjected to an uplift force, the soil located above the anchor will

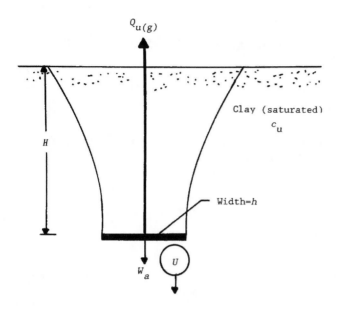

Figure 2-46   Horizontal anchor in saturated clay

be compressed and, at the same time, the soil below the anchor will be relieved of some stress. This will, in turn, result in an increase of the pore water pressure above the anchor accompanied by a decrease of pore water pressure below the anchor. The difference will result in a suction force. This suction force will increase the short-term uplift capacity of the anchor. Thus the uplift capacity can be given by the expression

$$Q_{u(g)} = Q_u + W_a + U \qquad\qquad (2\text{-}64)$$

where $Q_{u(g)}$ and Q_u = gross and net ultimate uplift capacity, respectively
W_a = effective weight of the anchor
U = suction force below the anchor

Very little is known at the present time about the magnitude of the suction force and its variation with depth and type of clay soil. However, for design purposes the suction force should be neglected and the net ultimate uplift capacity should be taken as

$$Q_u = Q_{u(g)} \quad W_a \qquad\qquad (2\text{-}65)$$

In the following subsections, the existing theories for estimation of the net uplift capacity Q_u will be summarized.

2.9.1   Theory of Vesic

In Section 2.3.6 it was shown that, for anchors embedded in sand ($c = 0$)

$$Q_v = A\gamma H F_q \qquad\qquad (2\text{-}48)$$

where A = area of the anchor plate

The preceding relation was derived by Vesic (1971) using the analogy of expansion of cavities.   In a similar manner, it can be shown that, in a c-ϕ soil

$$Q_u = A(\gamma H F_q + c F_c) \qquad\qquad (2\text{-}66)$$

where F_c = breakout factor
c = cohesion of the soil

For undrained condition, $\phi = 0$ and $c = c_u$.   It was shown in Tables 2-3 and 2-4 that, for $\phi = 0$, the value of F_q is equal to one.   Thus

$$Q_u = A(\gamma H + c_u F_c) \qquad\qquad (2\text{-}67)$$

Vesic (1971) presented the theoretical variation of the breakout factor F_c (for $\phi = 0$ condition) with the embedment ratio H/h, and these values are given in Table 2-6.   A plot of these same values of F_c against H/h is also shown in Figure 2-47.   Based on the laboratory model test results available at the present time, it appears that Vesic's theory gives a closer estimate only for shallow anchors embedded in softer clay.

Table 2-6   Variation of F_c ($\phi = 0$ condition)

	H/h				
Anchor type	0.5	1.0	1.5	2.5	5.0
Circular (diameter = h)	1.76	3.80	6.12	11.6	30.3
Strip ($h/B \approx 0$)	0.81	1.61	2.42	4.04	8.07

In general, the breakout factor increases with embedment ratio up to a maximum value and remains constant thereafter as shown in Figure 2-48.   The maximum value of $F_c = F_c^*$ is reached at $H/h = (H/h)_{cr}$.   Anchors located at H/h

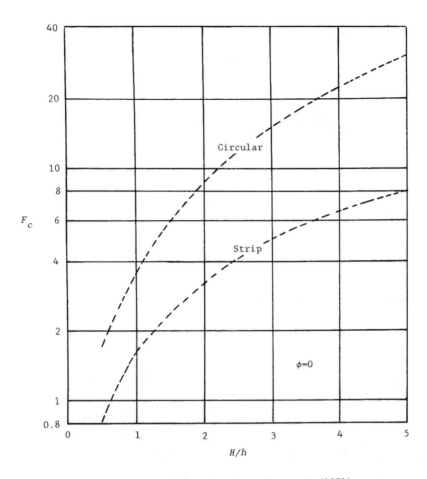

Figure 2-47   Variation of Vesic's breakout factor F_c (1971)

$> (H/h)_{cr}$ are referred to as deep anchors. For these anchors, at ultimate uplifting load, local shear failure in soil located around the anchor takes place. Anchors located at $H/h \leq (H/h)_{cr}$ are shallow anchors.

2.9.2   Theory of Meyerhof

Based on experimental results, Meyerhof (1973) proposed the following relationship

$$Q_u = A(\gamma H + F_c c_u)$$

For *circular* and *square* anchors

$$F_c = 1.2(\frac{H}{h}) \leq 9 \qquad\qquad (2\text{-}68)$$

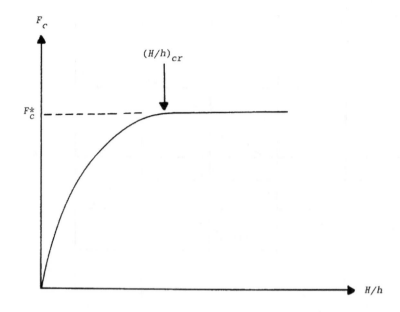

Figure 2-48   Nature of variation of F_c with H/h

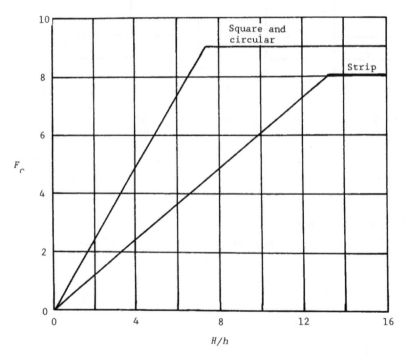

Figure 2-49   Variation of F_c with H/h [Equations (2-68) and (2-69)]

and, for *strip* anchors

$$F_c = 0.6(\frac{H}{h}) \leq 8 \qquad\qquad (2\text{-}69)$$

Equations (2-68) and (2-69) imply that, for *circular* and *square* anchors

$$(\frac{H}{h})_{cr} = \frac{9}{1.2} = 7.5 \qquad\qquad (2\text{-}70)$$

and, for *strip* anchors

$$(\frac{H}{h})_{cr} = \frac{8}{0.6} \approx 13.5 \qquad\qquad (2\text{-}71)$$

The breakout factor variations with embedment ratio according to Equations (2-68) and (2-69) are shown in Figure 2-49. Based on the experimental results it appears that Equations (2-68) and (2-69) are reasonable estimates for anchors embedded in stiff clay.

2.9.3 Theory of Das

Das (1978) compiled a number of laboratory model test results on *circular* anchors embedded in saturated clay with the undrained cohesion c_u varying from 5.18 kN/m^2 to about 172.5 kN/m^2 Figure 2-50 shows the average plots of F_c versus H/h obtained from these studies, along with the critical embedment ratios. The details relating to curves a, b, c, d and e shown in Figure 2-50 are given in Table 2-7

Table 2-7   Details for the Curves Shown in Figure 2-50

Curve No.	Reference	Year	c_u (kN/m^2)
a	Ali	1968	5.18
b	Kupferman	1971	6.9
c	Adams and Hayes	1967	10.35-13.8
d	Bhatnagar	1969	53.17
e	Adams and Hayes	1967	96.6-172.5

From Figure 2-50 it can be seen that, for shallow anchors

$$F_c \approx n(\frac{H}{h}) \leq 8 \text{ to } 9 \qquad\qquad (2\text{-}72)$$

where n = a constant

72

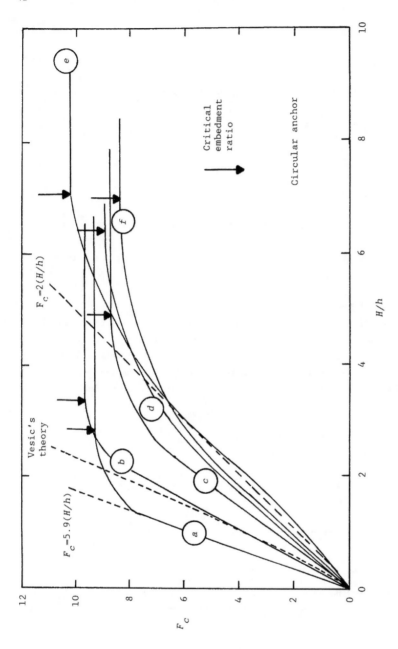

Figure 2-50 Variation of breakout factor with H/h for various experimental observations

The magnitude of n varies between 5.9 to 2.0 and is a function of the un-drained cohesion. Since n is a function of c_u and $F_c = F_c^*$ is about 8 to 9 in all cases, it is obvious that the critical embedment ratio $(H/h)_{cr}$ will be a function of c_u.

Das (1978) also reported some model test results conducted with *square* and *rectangular* anchors of width $h = 50.8$ mm. Based on these model test results, the variation of F_c with H/h is shown in Figure 2-51. Using the critical embedment ratios obtained from Figures 2-50 and 2-51, it was proposed that

$$(\frac{H}{h})_{cr-S} = 0.107c_u + 2.5 \leq 7 \tag{2-73}$$

where $(H/h)_{cr-S}$ = critical embedment ratio of *square* anchor (or *circular* anchor)

c_u = undrained cohesion in kN/m^2

A plot based on Equation (2-73) is shown in Figure 2-52. It was also observed by Das (1980) that

$$(\frac{H}{h})_{cr-R} = (\frac{H}{h})_{cr-S}[0.73 + 0.27(\frac{B}{h})] \leq 1.55(\frac{H}{h})_{cr-S} \tag{2-74}$$

where $(H/h)_{cr-R}$ = critical embedment ratio of rectangular anchors (Figure 2-53)

Based on these model test results, Das (1980) proposed an empirical pro-cedure to obtain the breakout factors for shallow and deep anchors. According to this procedure, α' and β' are two nondimensional factors defined as

$$\alpha' = \frac{\frac{H}{h}}{(\frac{H}{h})_{cr}} \tag{2-75}$$

and

$$\beta' = \frac{F_c}{F_c^*} \tag{2-76}$$

For a given anchor (that is, circular, square or rectangular), the critical embedment ratio can be calculated by using Equations (2-73) and (2-74). The magnitudes of F_c^* can be given by the following empirical relationship

$$F_{c-R}^* = 7.56 + 1.44(\frac{h}{B}) \tag{2-77}$$

74

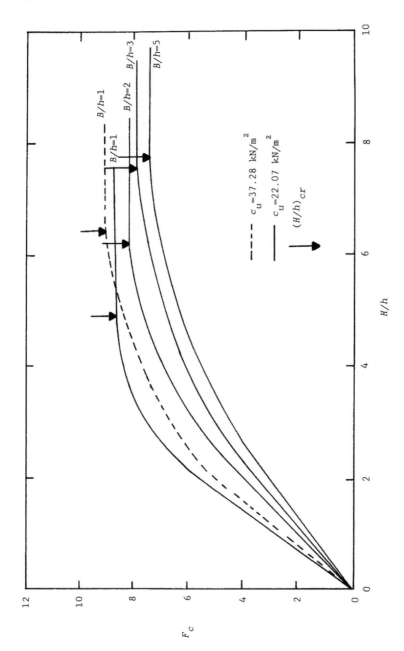

Figure 2-51 Model test results of Das (1978)--variation of F_c with H/h

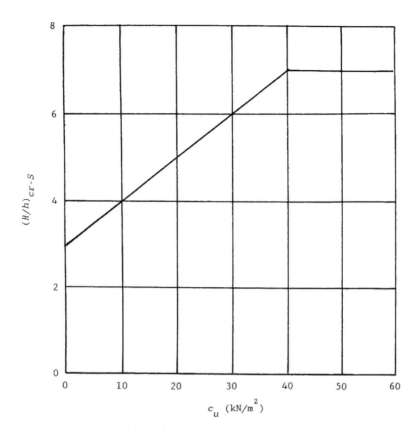

Figure 2-52 Plot of $(H/h)_{cr-S}$ versus c_u (kN/m^2)--Equation (2-73)

where F^*_{c-R} = breakout factor for deep *rectangular* anchor

It can be seen from Equation (2-77) that for *square* and *circular* anchors F^*_{c-R} is equal to 9. Using all the experimental curves shown in Figures 2-50 and 2-51, when the nondimensional breakout factor β' is plotted against the nondimensional embedment ratio α', they appear to fall in a rather narrow range as shown in Figure 2-54. The average plot of β' versus α' is also shown in Figure 2-54. Hence, following is a step-by-step procedure for estimation of the net ultimate uplift capacity.

1. Determine the representative value of the undrained cohesion c_u
2. Determine the critical embedment ratio using Equations (2-73) and (2-74).
3. Determine the H/h ratio for the anchor.

76

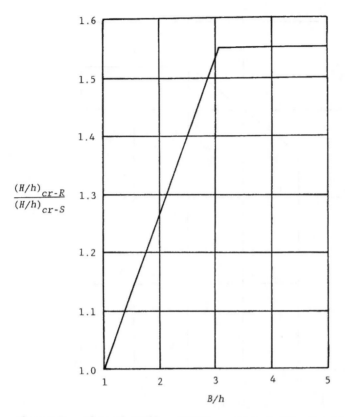

Figure 2-53 Plot of $(H/h)_{cr-R}/(H/h)_{cr-S}$ against B/h--Equation (2-74)

4. If $H/h > (H/h)_{cr}$ as determined in Step 2, it is a deep anchor.
   However if $H/h \leq (H/h)_{cr}$ it is a shallow anchor.

5. For $H/h > (H/h)_{cr}$

$$F_c = F_c^* = 7.56 + 1.44(\frac{h}{B})$$

Thus

$$Q_u = A\{[7.56 + 1.44(\frac{h}{B})]c_u + \gamma H\} \qquad (2-78)$$

where A = area of the anchor

6. For $H/h \leq (H/h)_{cr}$

$$Q_u = A(\beta'F_c^*c_u + \gamma H) = A\left\{\beta'[7.56 + 1.44(\frac{h}{B})]c_u + \gamma H\right\} \qquad (2-79)$$

The value of β' can be obtained from the average curve of Figure 2-54.

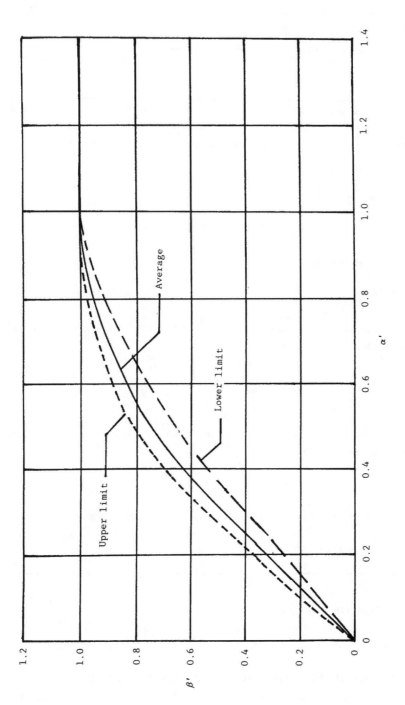

Figure 2-54 Plot of β' versus α' (after Das, 1980)

78

The procedure outlined above gives fairly good results in estimating the net ultimate holding capacity of anchors.

Example

2.6

A plate anchor measuring 0.4 m $\times$ 0.6 m is embedded at a depth of 1.8 m. The undrained cohesion of the clay is 42 kN/m^2 The saturated unit weight γ is 18.9 kN/m^3 Estimate the net ultimate uplift capacity.

Solution

From Equation (2-73)

$$(\frac{H}{h})_{cr-S} = 0.107c_u + 2.5 = (0.107)(42) + 2.5 \approx 7$$

Again, from Equations (2-74)

$$(\frac{H}{h})_{cr-R} = (\frac{H}{h})_{cr-S}[0.73 + 0.27(\frac{B}{h})] = (7)[0.73 + (0.27)(\frac{0.6}{0.4})] \approx 7.95$$

The actual embedment ratio is $H/h = 1.8/0.4 = 4.5$. Hence this is a shallow anchor.

$$\alpha' = \frac{\frac{H}{h}}{(\frac{H}{h})_{cr}} = \frac{4.5}{7.95} = 0.566$$

Referring to Figure 2-54, for $\alpha' = 0.566$ the magnitude of β' is 0.82. From Equation (2-79)

$$Q_u = A\{\beta'[7.56 + 1.44(\frac{h}{B})]c_u + \gamma H\}$$

$$= (0.4)(0.6)\{(0.82)[7.56 + (1.44)(\frac{0.4}{0.6})](42) + (18.9)(1.8)\}$$

$$= \textbf{78.6 kN}$$

2.10   FACTOR OF SAFETY

In most cases of anchor design, it is recommended that a factor of safety of 2 to 2.5 be used to arrive at the allowable net ultimate uplift capacity.

2.11   UPLIFT CAPACITY OF ANCHORS IN LAYERED SOIL

The uplift capacity of anchors embedded in a saturated clay layer overlain by a compact sand deposit was studied by Stewart (1985) using laboratory model tests. The basic conclusions of this study can qualitatively be summarized by referring to Figure 2-55. In Figure 2-55a a plate anchor is embedded in a saturated clay at a depth $H = H_1$. When subjected to an uplifting load, the nature of the plot of the net load Q versus anchor uplift Δ will be of the type shown by curve a in Figure 2-56. If H_1 is relatively small, then the failure surface will extend to the top of the clay layer indicating shallow anchor condition. If a layer of dense sand is now placed on the clay layer, the total thickness of the soil above the anchor will be equal to H_2 (Figure 2-55b). For this condition, the nature of the Q versus Δ plot will be as shown by curve b in Figure 2-56. For this condition the sand acts as a surcharge on the clay layer and increases the net ultimate uplift capacity. If the thickness of the clay layer is gradually increased, depending on the relative value of c_u, the angle of friction of sand, γ_{clay} and γ_{sand}, there will be a condition when the anchor will behave like a deep anchor located in clay. For this condition, let the thickness of the sand and clay above the anchor be equal to H_3 as shown in Figure 2-55c. Curve c in Figure 2-56 represents the Q versus Δ plot for this condition. If the thickness of the sand layer is further increased (Figure 2-55d) and an uplifting load is applied to the anchor, the load-displacement plot will follow the path shown by curve d (Figure 2-56) which is the same path as shown by curve c in Figure 2-56. However, if sufficient upward anchor displacement is allowed such that the anchor reaches the top of the sand (Figure 2-55e), then the load resistance increases again and follows the path shown by curve e in Figure 2-56. Based on this, we can draw the following conclusions:

1. The sand overlay can significantly increase the net ultimate uplift capacity.

2. The net ultimate uplift capacity is composed of two parts, namely

$$Q_u = Q_{u(clay)} + Q_{u(sand)} \qquad (2\text{-}80)$$

where $Q_{u(clay)}$ = clay component
$Q_{u(sand)}$ = sand component

The magnitude of $Q_{u(clay)}$ increases with the H/h ratio up to a maximum value at $H/h = H_3/h$ (Figure 2-55c). A further increase of H/h has no effect on the magnitude of $Q_{u(clay)}$. The sand component, $Q_{u(sand)}$, is mobilized only when the anchor plate punches through the clay layer and reaches the sand-clay interface.

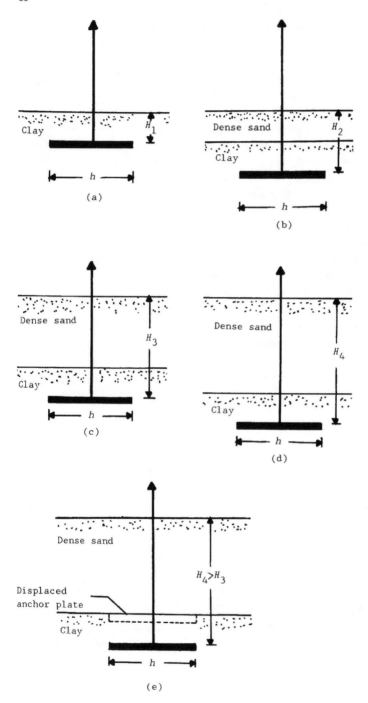

Figure 2-55   Plate anchor in saturated clay overlain by dense sand

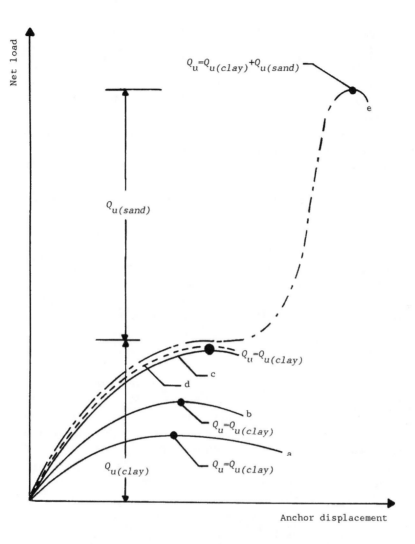

Figure 2-56   Nature of net load versus anchor displacement plots
for plate anchor in clay overlain by dense sand

REFERENCES

Adams, J.I. and Hayes, D.C., 1967. The uplift capacity of shallow foundations. *Ontario Hydro Res. Qtrly.*, 19(1):1-13.

Ali, M., 1968. Pullout resistance of anchor plates in soft Bentonite clay.
M.S. Thesis, Duke University, Durham, N.C., USA.

Andreadis, A., Harvey, R.C. and Burley, E., 1978. Embedment anchors subjected
to repeated loading. *J. Geotech. Engrg. Div.*, ASCE, 104(7):1033-1036.

Baker, W.H. and Kondner, R.L., 1966. Pullout load capacity of a circular earth anchor buried in sand. *Highway Res. Rec. 108*, National Academy of Sciences, Washington, D.C., 1-10.

Balla, A., 1961. The resistance to breaking-out of mushroom foundations for pylons. *Proc., V Intl. Conf. Soil Mech. Found. Engrg.*, Paris, France, 1:569-576.

Bhatnagar, R.S., 1969. Pullout resistance of anchors in silty clay. M.S. Thesis, Duke University, Durham, N.C., USA.

Caquot, A. and Kerisel, L., 1949. *Traite de mecanique des sols*. Gauthier-Villars, Paris, France.

Das, B.M., 1980. A procedure for estimation of ultimate uplift capacity of foundations in clay. *Soils and Found.*, Japan, 20(1):77-82.

Das, B.M., 1978. Model tests for uplift capacity of foundations in clay. *Soils and Found.*, Japan, 18(2):17-24.

Das, B.M. and Jin-Kaun, Y., 1987. Uplift capacity of model group anchors in sand. *Found. for Transmission Towers*, Geotech. Spec. Pub. 8, ASCE:57-71.

Das, B.M. and Jones, A.D., 1982. Uplift capacity of rectangular foundations in sand. *Trans. Res. Rec. 884*, National Academy of Sciences, Washington, D.C., 54-58.

Das, B.M. and Seeley, G.R., 1975a. Breakout resistance of horizontal anchors. *J. Geotech. Engrg. Div.*, ASCE, 101(9):999-1003.

Das, B.M. and Seeley, G.R., 1975b. Inclined load resistance of anchors in sand. *J. Geotech. Engrg. Div.*, ASCE, 101(9):995-998.

Das, B.M. and Puri, V.K., 1989. Load-displacement relationship for horizontal rectangular anchors in sand. *Proc., IV Intl. Conf. Civil and Structural Engr. Computing*, London, Civil-COMP Press, Edinburgh, 2:207-212.

Downs, D.I. and Chieurzzi, R., 1966. Transmission tower foundations. *J. Power Div.*, ASCE, 88(2):91-114.

Esquivel-Diaz, R.F., 1967. Pullout resistance of deeply buried anchors in sand. M.S. Thesis, Duke University, Durham, N.C., USA.

Ireland, H.O., 1963. Discussion, Uplift resistance of transmission tower footing. *J. Power Div.*, ASCE, 89(1):115-118.

Kulhawy, F.H., Trautmann, C.H. and Nicolaides, C.N., 1987. Spread foundations in uplift: Experimental study. *Found. for Transmission Towers*, Geotech. Spec. Pub. 8, ASCE:110-127.

Kupferman, M., 1971. The vertical holding capacity of marine anchors in clay subjected to static and cyclic loading. M.S. Thesis, University of Massachusetts, Amherst, USA.

Mariupol'skii, L.G., 1965. The bearing capacity of anchor foundations (English translation). *Soil Mech. Found. Engrg.* (in Russian):26-37

Mors,, H., 1959 The behavior of mast foundations subjected to tensile forces. *Bautechik*, 36(10):367-378.

Meyerhof, G.G., 1973. Uplift resistance of inclined anchors and piles. *Proc.*, VIII Intl. Conf. Soil Mech. Found. Engrg., Moscow, USSR, 2.1:167-172.

Meyerhof, G.G. and Adams, J.I., 1968. The ultimate uplift capacity of foundations. *Can. Geotech J.*, 5(4):225-244.

Saeedy, H.S., 1987 Stability of circular vertical earth anchors. *Can. Geotech. J.*, 24(3):452-456.

Stewart, W., 1985. Uplift capacity of circular plate anchors in sand. *Can. Geotech J.*, 22(4):589-592.

Sutherland, H.B., 1965. Model studies for shaft raising through cohesionless soils. *Proc.*, VI Intl. Conf. Soil Mech. Found. Engrg., Montreal, Canada, 2:410-413.

Veesaert, C.J. and Clemence, S.P., 1977. Dynamic pullout resistance of anchors. *Proc.*, Intl. Symposium on Soil-Structure Interaction, Rourkee, India, 1:389-397.

Vesic, A.S., 1971. Breakout resistance of objects embedded in ocean bottom. *J. Soil Mech. Found. Div.*, ASCE, 97(9):1183-1205.

Vesic, A.S., 1965. Cratering by explosives as an earth pressure problem. *Proc.*, VI Intl. Conf. Soil Mech. Found. Engrg., Montreal, Canada, 2:427-431.

Chapter 3
VERTICAL PLATE ANCHORS

3.1   INTRODUCTION

The use of vertical anchor plates to resist horizontal loading in the construction of sheet pile walls has been discussed in Chapter 1 (Figure 1-4).
Inadequate design of anchors has been the cause of failure of many sheet pile
walls.   Sheet pile walls are flexible structures and, due to the outward bulg-
ing of these walls, the lateral earth pressure produced is quite different
than that calculated for rigid structures using the classical Rankine or
Coulomb earth pressure theories.   In conducting laboratory measurements, Rowe
(1952) showed that the bending moment to which an anchored sheet pile wall is
subjected can be substantially reduced when the anchor movement is less than
about 0.1% of the height of the wall.   The movement of 0.1% of the anchor includes the elongation of the tie rod connecting the vertical plate anchors and
the wall.   Hence it is important to make proper estimation of the ultimate and
allowable holding capacities of plate anchors and also the corresponding dis-
placements.   Vertical plate anchors can also be used at pressure pipeline
bends, at the base of retaining walls to resist sliding (Figure 3-1), and also
where it is necessary to control thermal stresses.

Figure 3-2 shows the geometric parameters of a vertical anchor plate.
The height and width of the anchor plate are h and B, respectively.   The depth
of embedment of the anchor plate is H (that is, the distance from the ground
surface to the bottom of the plate).   In most practical cases, the anchor can
be considered as a strip anchor (two-dimensional plane strain case) if the B/h
ratio is greater than about 6.

The holding capacity of an anchor is primarily derived from the passive
force imposed by the soil in front of the anchor slab.   If the embedment ratio
H/h of the anchor is relatively small, at ultimate pullout load on the anchor
the passive failure surface developed in soil in front of the anchor will intersect the ground surface.   This is referred to (as in the case of horizontal
anchors; Chapter 2) as *shallow anchor condition*.   Figure 3-3 shows the failure
surface in front of a shallow square plate anchor (that is, $h = B$) embedded in
sand as observed by Hueckel (1957).   At greater embedment ratios

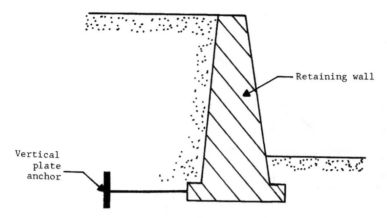

Figure 3-1   Vertical plate anchor at the base of a retaining wall to resist sliding

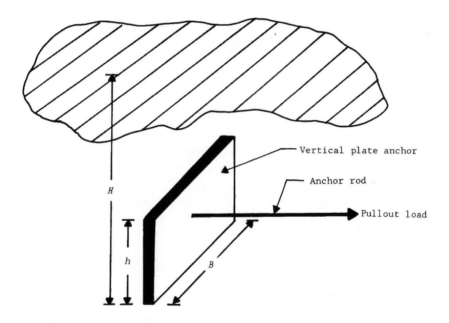

Figure 3-2   Geometric parameters of a vertical plate anchor

the *local shear failure* in soil will take place at ultimate load, and these anchors are called *deep anchors*. So the ultimate holding capacity Q_u is a function of several parameters,

    a.   *H/h* ratio;

    b.   width-to-height ratio, *B/h*;

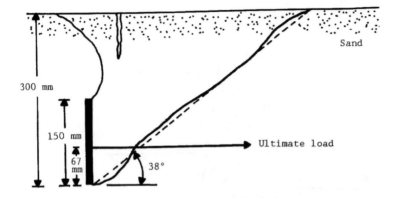

Figure 3-3   Failure surface in front of a square anchor slab (150 mm × 150 mm) embedded in sand at $H/h = 2$ as observed by Hueckel (1957) (*Note*: Soil friction angle $\phi = 34°$)

c.   shear strength parameters of the soil (soil friction angle, ϕ, and cohesion, c); and

d.   the angle of friction at the anchor-soil interface, δ.

In the following sections the effect of the above factors on the holding capacity of anchors will be analyzed in detail. As in the case of horizontal plate anchors (Chapter 2), this chapter has been divided into two major parts: (a) behavior of anchors in sand and (b) behavior of anchors in clay (undrained condition). It is important to note that, for vertical anchors, the gross ultimate holding capacity $Q_{u(g)}$ is equal to the net ultimate holding capacity Q_u.

ANCHORS IN SAND

3.2   ULTIMATE HOLDING CAPACITY FROM RANKINE'S THEORY

One of the earlier methods for estimation of the holding capacity of vertical anchors was by using the theory of *Rankine's lateral earth pressure* (Teng, 1962). Figure 3-4a shows a vertical strip anchor embedded in a granular soil, at a relatively shallow depth. The relatively *shallow depth condition* refers to the case where $h/H \leq 1/3$ to $1/2$. Assuming the Rankine state exists, the failure surface in soil around the anchor at ultimate load is also shown in Figure 3-4a.

According to this procedure, for a strip anchor the ultimate holding capacity *per unit width* (that is, at right angles to the cross section shown in Figure 3-4b) can be given as

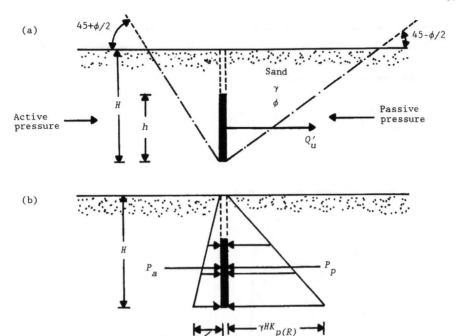

Figure 3-4   Ultimate holding capacity of strip vertical anchor as derived by Teng (1962)

$$Q'_u = P_p \quad P_a \tag{3-1}$$

where Q'_u = ultimate holding capacity *per unit width*

P_p = passive force in front of the anchor *per unit width* (Figure 3-4b)

P_a = active force at the back of the anchor *per unit width* (Figure 3-4b)

The relationships for P_p and P_a are as follows

$$\left. \begin{aligned} P_p &= \frac{1}{2}\gamma H^2 \tan^2(45 + \frac{\phi}{2}) \\ P_a &= \frac{1}{2}\gamma H^2 \tan^2(45 \quad \frac{\phi}{2}) \end{aligned} \right\} \tag{3-2}$$

where

γ = unit weight of soil

ϕ = soil friction angle

$\tan^2(45 + \phi/2) = K_{p(R)}$ = Rankine passive earth pressure coefficient

$\tan^2(45 \quad \phi/2) = K_{a(R)}$ = Rankine active earth pressure coefficient

88

For anchors with a limited width B, the frictional resistance developed along the vertical faces of the failure surface must be taken into account (Figure 3-5). Following the procedure of Teng (1962), the *total earth pressure normal* to *abc* and *def* is

$$N = 2 \int_{o}^{H} (\frac{H - z}{H})[H\sqrt{K_{p(R)}} + H\sqrt{K_{a(R)}}](dz)(\gamma K_o)$$

$$= \frac{1}{3}K_o\gamma[\sqrt{K_{p(R)}} + \sqrt{K_{a(R)}}]H^3 \qquad (3\text{-}3a)$$

where K_o = earth pressure coefficient at rest ≈ 0.4

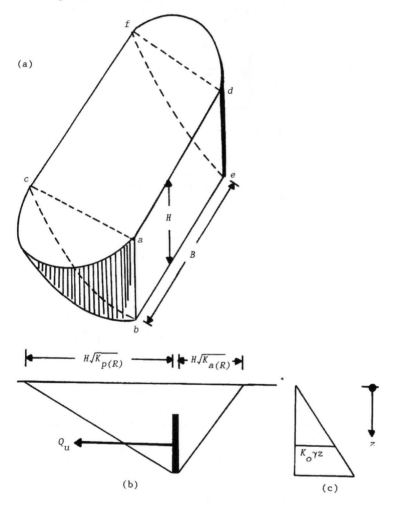

Figure 3-5   Frictional resistance developed along the vertical faces of the failure surface--Teng's method

Hence, the total frictional resistance at the ends is

$$F = N \tan \phi = \frac{1}{3} K_o [\sqrt{K_{p(R)}} + \sqrt{K_{a(R)}}] H^3 \tan \phi \qquad (3\text{-}3b)$$

So, the ultimate holding capacity can be given as

$$Q_u = Q_u' B + F = B(P_p - P_a) + \frac{1}{3} K_o [\sqrt{K_{p(R)}} + \sqrt{K_{a(R)}}] H^3 \tan \phi \qquad (3\text{-}4)$$

Example

3.1

For a vertical anchor, the following values are given: $h = 2$ ft, $B = 5$ ft, $H = 4$ ft, $\gamma = 105$ lb/ft^3, $\phi = 32°$ Determine the ultimate holding capacity, Q_u

Solution

$$K_{p(R)} = \tan^2(45 + \frac{\phi}{2}) = \tan^2(45 + \frac{32}{2}) = 3.25$$

$$K_{a(R)} = \tan^2(45 - \frac{\phi}{2}) = \tan^2(45 - \frac{32}{2}) = 0.307$$

$$P_p = \frac{1}{2}\gamma H^2 K_{p(R)} = (\frac{1}{2})(105)(4)^2(3.25) = 2730 \text{ lb/ft}$$

$$P_a = \frac{1}{2}\gamma H^2 K_{a(R)} = (\frac{1}{2})(105)(4)^2(0.307) = 257.9 \text{ lb/ft}$$

$$F_3 = \frac{1}{3} K_o \gamma [\sqrt{K_{p(R)}} + \sqrt{K_{a(R)}}] H^3 \tan \phi$$

$$= (\frac{1}{3})(0.4)(105)(\sqrt{3.25} + \sqrt{0.307})(4)^3 \tan 32° = 699.25 \text{ lb}$$

So

$$Q_u = B(P_p - P_a) + F = (5)(2730 - 257.9) + 699.25 \approx \mathbf{13{,}060 \text{ lb}}$$

3.3   RECENT DEVELOPMENTS ON SHALLOW VERTICAL ANCHORS IN SAND

During the last twenty years or so, a number of theoretical and experimental studies were conducted to define the actual failure surface in soil around the anchor at ultimate load and to determine the ultimate holding capacity of vertical anchors in sand. Some of these recent developments will be discussed in the following subsections.

3.3.1   Analysis of Ovesen and Stromann

In 1964, Ovesen reported the results of several model tests conducted for shallow anchors in sand at the Danish Geotechnical Institute. The method of

90

analysis developed in this subsection (Ovesen and Stromann, 1972) is primarily based on those model tests and is based on the following concepts.

1. Determination of the holding capacity per unit width of a continuous anchor plate, $Q'_{u(B)}$, of height H as shown in Figure 3-6a. This is known as the *basic case*.

2. Estimation of the holding capacity per unit width of an anchor whose height is h (Figure 3-6b) and has an embedment depth of H ($h \leq H$). This is known as the *strip case*.

3. Estimation of the holding capacity of an *anchor with limited width-to-height ratio* (B/h; Figure 3-6c).

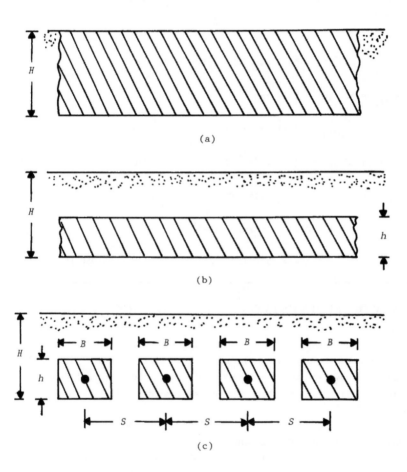

(a)

(b)

(c)

Figure 3-6   Ovesen and Stromann's analysis: (a) basic case;   (b) strip case;   (c) actual case

Basic Case

Figure 3-7 shows the basic case for a vertical anchor embedded in sand. The assumed failure surface in soil for *translation* of the anchor at ultimate load $Q'_{u(B)}$ (load per unit width) is also shown in this figure. For rough anchor surface, P_a is the active force per unit width. The horizontal and vertical components of P_a can be given as

$$P_{a(H)} = P_a \cos \phi \qquad (3-5)$$

$$P_{a(V)} = P_a \sin \phi \qquad (3-6)$$

where $P_{a(H)}, \ P_{a(V)}$ = horizontal and vertical components of P_a, respectively
ϕ = soil friction angle

The passive failure surface in front of the anchor slab consists of (a) straight rupture line BC, (b) Prandtl radial shear zone ACD, and (c) Rankine passive zone ADE. (Note: Angles EAD and AED are both equal to $45 \quad \phi/2$.)

The horizontal and vertical components of the passive force P_p are

$$P_{p(H)} = \frac{1}{2}\gamma H^2 K_{pH} \qquad (3-7)$$

$$P_{p(V)} = \frac{1}{2}\gamma H^2 K_{pH} \tan \delta \qquad (3-8)$$

where γ = unit weight of the soil
K_{pH} = horizontal component of the passive earth pressure coefficient
δ = anchor-soil friction angle

For vertical equilibrium

$$P_{a(V)} + W = P_{p(V)} \qquad (3-9)$$

where W = weight of anchor per unit width

From Equations (3-8) and (3-9)

$$K_{pH}\tan \delta = \frac{P_{a(V)} + W}{\frac{1}{2}\gamma H^2} \qquad (3-10)$$

Figure 3-8 shows the variation of $K_{pH}\tan\delta$ and ϕ from which K_{pH} can be estimated. Now, for horizontal equilibrium

$$Q'_{u(B)} = P_{p(H)} \qquad P_{a(H)} = \frac{1}{2}\gamma H^2 K_{pH} \qquad P_{a(H)} \qquad (3-11)$$

92

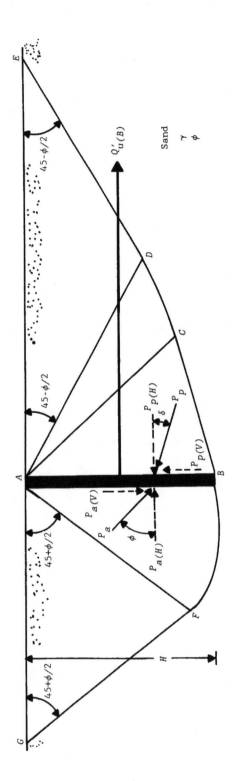

Figure 3-7 Basic case--failure surface at ultimate load

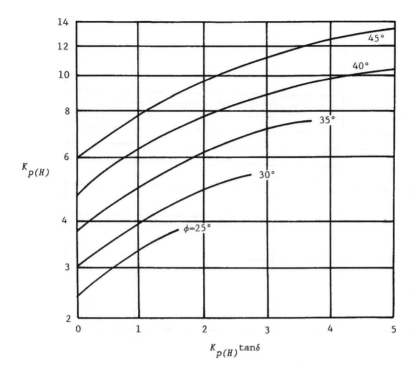

Figure 3-8 Variation of $K_{p(H)}$ with $K_{p(H)}\tan\delta$ and ϕ--Ovesen and Stromann's analysis

The magnitudes of $P_{a(V)}$ and $P_{a(H)}$ can be determined using any ordinary active earth pressure theory. Figure 3-9 shows the variation of the active earth pressure coefficient K_a according to Caquot and Kerisel (1949). Note that

$$P_a = \tfrac{1}{2}\gamma H^2 K_a \tag{3-12}$$

Strip Case
Based on the experimental evidence of Ovesen (1964), the ultimate holding capacity of a *strip anchor* can be given as

$$Q_u' = R_{ov}Q_{u(B)}' \tag{3-13}$$

where Q_u' = ultimate holding capacity per unit width for strip anchor

The variation of R_{ov} with the ratio h/H is shown in Figure 3-10. Note that

$$R_{ov} = \frac{C_{ov} + 1}{C_{ov} + \frac{H}{h}} \tag{3-14}$$

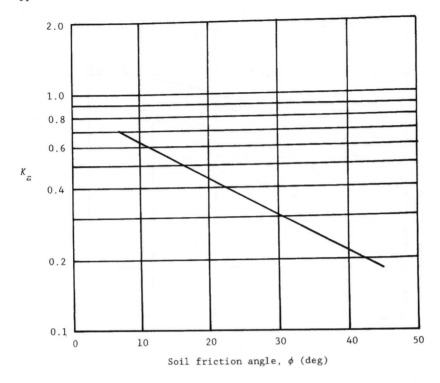

Figure 3-9   Variation of K_a with ϕ

where C_{ov} = $\begin{cases} 19 \text{ for dense sand} \\ 14 \text{ for loose sand} \end{cases}$

Equation (3-14) was developed by Dickin and Leung (1985) from Figure 3-10.

Anchor with Limited B/h Ratio

When an anchor has a limited width to height ratio (B/h), the failure surface in soil will be three-dimensional as shown in Figure 3-5. Hence, the ultimate holding capacity of an anchor, Q_u, can be given as

$$Q_u = Q_u'B + F$$

where F = frictional resistance derived from the sides of the three-dimensional failure surface

However if a number of vertical anchors are used in a row, depending on the S/B ratio (S = center-to-center spacing of the anchor as shown in Figure 3-11), the failure surface may overlap. In that case

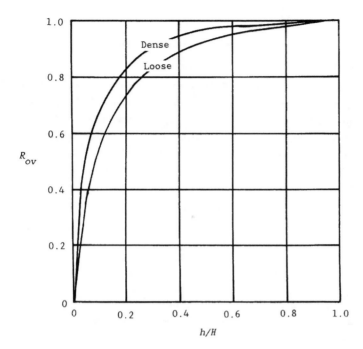

Figure 3-10 Variation of R_{ov} with H/h based on Ovesen and Stromann's analysis

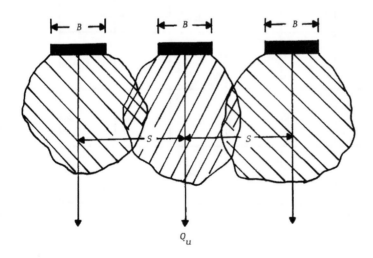

Figure 3-11 Overlapping of failure surface in soil in front of a row of vertical anchors

96

$$Q_u = Q_u'B + F'$$

where F' = frictional resistance $\leq F$

Hueckel (1957) conducted a number of laboratory model tests on three square anchors (Figure 3-11) to determine the S/B ratio at which $F = F'$, and these results are shown in Figure 3-12. Note that Q_{ug} is the notation for the holding capacity of the anchor group. In this case, the group consists of three anchors, each measuring 100 mm × 100 mm. From this figure, it can can be seen that, at $S/B \approx 3$ to 4, the effect of interference practically disappears.

Ovesen and Stromann (1972) expressed Q_u as

$$Q_v = Q_u'B_e \tag{3-15}$$

where B_e = equivalent width $\leq B$ (Figure 3-13)

The variation of B_e can be obtained from Figure 3-14.

In the case of a single anchor, that is $S = \infty$, we can also write that

$$Q_u = Q_u'BS_f \tag{3-16}$$

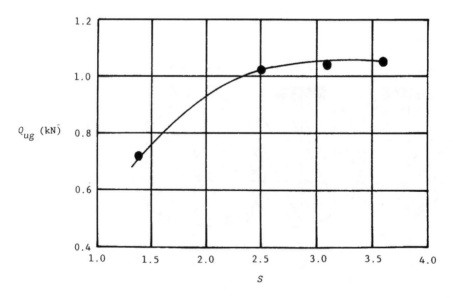

Figure 3-12   Variation of the ultimate group capacity with center-to-center spacing of anchor as observed by Hueckel (1957) (*Note*: $B = h = 100$ mm; $H/h = 2$; number of anchors = 3; $\phi = 36°$)

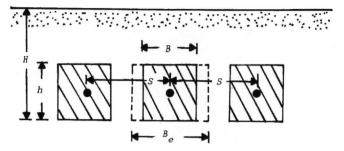

Figure 3-13   Definition of equivalent width

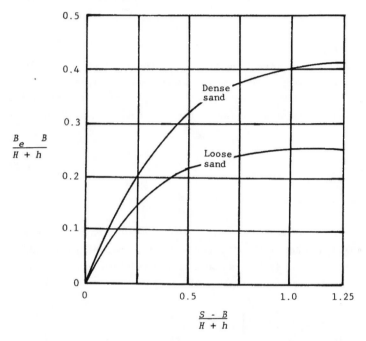

Figure 3-14   Variation of $(B_e-B)/(H+h)$ with $(S-B)/(H+h)$ according to Ovesen and Stromann's analysis (1972)

where $S_f = B_e/B$ = shape factor

From Figure 3-14, it can be shown that, with $S = \infty$ (also see Dickin and Leung, 1983)

$$S_f = 0.42 \left(\dfrac{\dfrac{H}{h} + 1}{\dfrac{B}{h}} \right) + 1 \quad \text{(for dense sand)} \tag{3-17}$$

and

$$S_f = 0.26 \left[\frac{\frac{H}{h} + 1}{\frac{B}{h}} \right] + 1 \quad \text{(for loose sand)} \tag{3-18}$$

Hence, for single anchors with limited width-to-height ratio, combining Equations (3-11), (3-14), (3-17), and (3-18), we obtain

$$Q_u = B[\frac{1}{2}\gamma H^2 K_{pH} \quad P_{a(H)}] \left[\frac{C_{ov} + 1}{C_{ov} + \frac{H}{h}} \right] \left[F\left[\frac{\frac{H}{h} + 1}{\frac{B}{h}} \right] + 1 \right] \tag{3-19}$$

where for dense sand, $C_{ov} = 19$ and $F = 0.42$; and, for loose sand, $C_{ov} = 14$ and $F = 0.26$.

3.3.2 Analysis of Meyerhof

Meyerhof (1973) took the passive and active earth pressure coefficients proposed by Caquot and Kerisel (1949) and Sokolovskii (1965) into consideration and proposed the following simple relationship for the ultimate holding capacity per unit width (Q'_u) of a continuous (strip) anchor. (See Figure 3.6b for notations.)

$$Q'_u = \frac{1}{2}\gamma H^2 K_b \tag{3-20}$$

where K_b = pullout coefficient

The variation of K_b is shown in Figure 3-15. It is the opinion of the author that, for a single anchor with a limited width-to-height ratio (B/h), Equations (3-14), (3-17), and (3-18) can be incorporated into Equation (3-20) to determine the ultimate holding capacity, or

$$Q_u = B(\frac{1}{2}\gamma H^2 K_b) \left[\frac{C_{ov} + 1}{C_{ov} + \frac{H}{h}} \right] \left[F\left[\frac{\frac{H}{h} + 1}{\frac{B}{h}} \right] + 1 \right] \tag{3-21}$$

The values of C_{ov} and F are given along with Equation (3-19)

3.3.3 Analysis of Biarez, Boucraut and Negre

Biarez, Boucraut, and Negre (1965) presented calculation methods for limiting equilibrium of vertical anchor piles subjected to translation and rotation. This analysis showed that, at an embedment ratio of $H/h < 4$, the ultimate holding capacity is a function of the weight and roughness, or

$$Q_u = f(W_a, \delta)$$

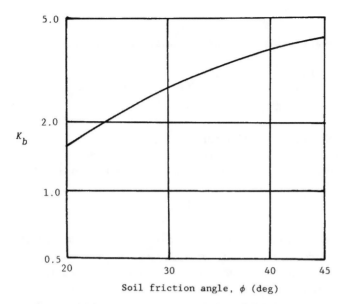

Figure 3-15 Meyerhof's pullout coefficient (1973)

where W_a = weight of anchor
 δ = anchor-soil friction angle

Dickin and Leung (1985) indicated that the analysis of Biarez et al. gave good results at embedment ratios of $4 < H/h < 7$. With the assumption of $\delta = 0$, the original equation of Biarez et al. (1965) can be conservatively expressed in a simplified form (Dickin and Leung, 1985)

$$Q'_u = \gamma h^2 \left\{ [K_{p(R)} \quad K_{a(R)}](\frac{H}{h} \quad \frac{1}{2}) + \left[\frac{K_{p(R)} \sin 2\phi}{2\tan(45 + \frac{\phi}{2})} \right](\frac{H}{h} \quad 1)^2 \right\}$$

(for strip anchor)   (3-22)

The preceding relationship can be expressed in a nondimensional form as

$$F_q = \frac{Q'_u}{\gamma h H} = \frac{h}{H} \left\{ [K_{p(R)} \quad K_{a(R)}](\frac{h}{H} \quad \frac{1}{2}) + \left[\frac{K_{p(R)} \sin 2\phi}{2\tan(45 + \frac{\phi}{2})} \right](\frac{H}{h} \quad 1)^2 \right\}$$ (3-23)

where F_q = breakout factor (similar to those given in Chapter 2). In the preceding two equations, $K_{a(R)}$ and $K_{p(R)}$ are Rankine active and passive earth pressure coefficients, respectively.

In a similar manner, the ultimate resistance of a shallow single anchor having dimensions of $B \times h$ can be expressed in a simplified nondimensional form as (Dickin and Leung, 1985)

$$F_q = \frac{Q_u}{\gamma(hB)H} = F_{q(strip)} + \phi(\tfrac{h}{B})(\tfrac{h}{H})[\sqrt{K_{p(R)}} \quad \sqrt{K_{a(R)}}](\tfrac{H}{h} \quad \tfrac{2}{3})$$

Eq. (3.23)

$$+ \tfrac{1}{2}(1 + \phi)(\tfrac{h}{B})(\tfrac{h}{H})K_{p(R)}\sin 2\phi(\tfrac{H}{h} \quad 1) \tag{3-24}$$

3.3.4 Analysis of Neely, Stuart and Graham

Neely, Stuart, and Graham (1973) predicted the holding capacity of vertical strip anchors using the *stress characteristics analysis* (Sokolovskii, 1965). The theoretical study consisted of developing the so-called force coefficient $M_{\gamma q}$ by two methods:

1. *Surcharge Method.* According to this method, it is assumed that the soil located above the top of the anchor can be taken as a simple surcharge of $q = \gamma(H \quad h)$. As shown in Figure 3-16, the failure surface in soil consists of an arc of a logarithmic spiral, *AC*, and a straight line, *CD*. Note that the zone *OCD* is a Rankine passive zone.

2. *Equivalent Free Surface Method.* The assumed failure surface in soil, *ACD* (Figure 3-17), is an arc of a logarithmic spiral with the center at *O*. *OD* is a straight line which is the *equivalent free surface*. The concept of the equivalent free surface is based on that developed by Meyerhof (1951) in the process of predicting the ultimate bearing capacity of foundations. Note that along the equivalent free surface, *OD*, the shear stress τ can be expressed as

$$\tau = m\sigma \tan \phi \tag{3-25}$$

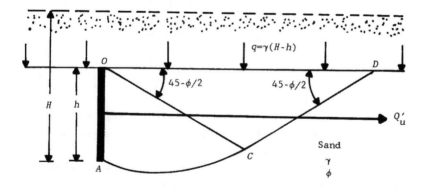

Figure 3-16 Surcharge method of analysis

101

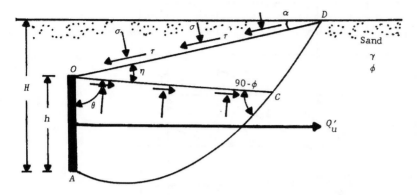

Figure 3-17   Failure mechanism assumed by Neely et al. (1973) for
analysis by the equivalent free surface method

where σ = effective normal stress

ϕ = soil friction angle

m = mobilization factor

The magnitude of m may vary between zero and one. When the value of
m is less than one, the failure mechanism includes a mixed shear
zone OCD in which the full shearing resistance of the soil becomes
mobilized.

In the assumed failure mechanism shown in Figure 3-17, note that

$$\frac{H}{h} = 1 + \frac{\sin \alpha \cdot \cos \phi \; e^{\theta \cdot \tan \phi}}{\cos(\phi + \eta)} \tag{3-26}$$

A nondimensional term $M_{\gamma q}$ (force coefficient) used earlier in this section may now be defined as

$$M_{\gamma q} = \frac{Q'_u}{\gamma h^2} \tag{3-27}$$

where Q'_u = ultimate holding capacity *per unit width of a strip anchor*

The relationship between the nondimensional force coefficient [Equation
(3-27)] and the breakout factor [Equation (3-23)] can be given as

$$F_q = M_{\gamma q}(\frac{h}{H}) \tag{3-28}$$

Figures 3-18 and 3-19 show the variation of $M_{\gamma q}$ for strip anchors determined by the surcharge method and the equivalent free surface method,
respectively. Note that, among other factors, the magnitude of $M_{\gamma q}$ based on

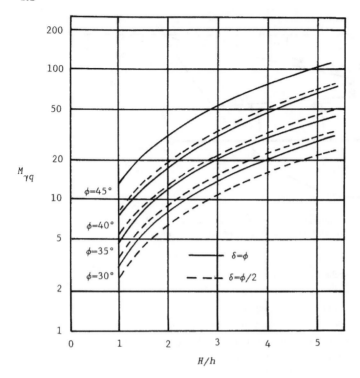

Figure 3-18 Variation of $M_{\gamma q}$ with H/h based on the surcharge method of analysis of Neely et al. (1973)

the surcharge method is a function of δ/ϕ (δ = soil-anchor friction angle). Similarly, the force coefficient based on the equivalent free surface method is a function of the mobilization factor, m. Neely et al. (1973) recommended the use of the analysis based on the equivalent free surface method.

As shown in Equation (3-16), the force coefficient (or the ultimate load) for a single anchor with a limited B/h ratio can be obtained by incorporating a nondimensional shape factor (S_f). In a similar manner

$$M_{\gamma q} = \frac{Q_u}{\gamma B h^2} = M_{\gamma q(strip)} S_f \qquad (3\text{-}29)$$

or

$$Q_u = (\gamma B h^2) M_{\gamma q(strip)} S_f \qquad (3\text{-}30)$$

The shape factor S_f given in Equations (3-29) and (3-30) was determined by Neely et al. (1973) from laboratory tests as

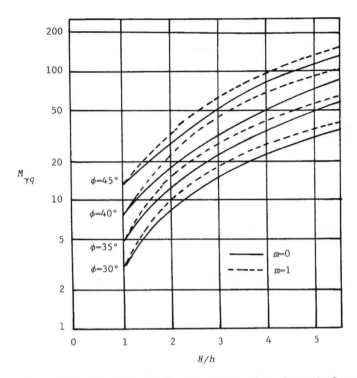

Figure 3-19   Variation of $M_{\gamma q}$ with H/h based on the equivalent free
surface method of analysis of Neely et al. (1973)

$$S_f = \frac{(\frac{Q_u}{B})_{rectangular}}{Q'_{u(strip)}} \qquad (3\text{-}31)$$

The variation of S_f, which is a function of H/h and B/h, as defined by
Equation (3-31) is shown in Figure 3-20. One important thing that needs to be
pointed out here is that Neely et al. (1973) assumed that an anchor with $B/h \geq$
5 should be treated as a strip anchor, or

$$Q'_{u(strip)} \approx Q'_{u(B/h=5)}$$

Das (1975) used the results of Neely et al. (1973) to express the ultimate holding capacity of *square* anchors ($B = h$) as

$$Q_u = C\gamma(\frac{H}{h})^n h^3 \qquad (3\text{-}32)$$

where $C = f(\phi)$
      $n = f(m)$

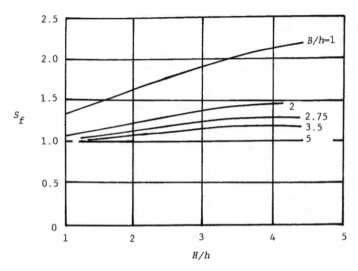

Figure 3-20   Variation of S_f [Equation (3-31)] (after Neely et al., 1973)

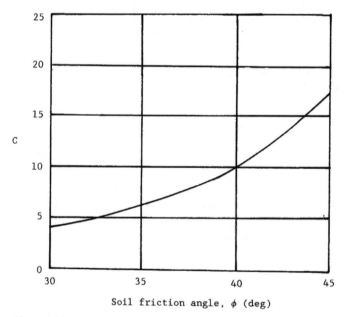

Figure 3-21   Variation of C with ϕ [Equation (3-32)]

The variation of C with the soil friction angle ϕ is shown in Figure 3-21. The magnitude of n is equal to 1.7 for mobilization factor $m = 0$, and $n = 1.9$ for $m = 1$. So based on a number of experimental results on small-scale laboratory model tests (Figure 3-22), Das (1975) suggested that the average value of n be taken as 1.8, or

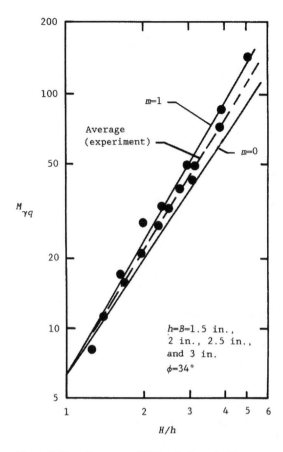

Figure 3-22   Force coefficient for shallow square anchors (based on the experimental results of Das, 1975)

$$Q_{u(square)} = C\gamma(\frac{H}{h})^{1 \cdot 8}h^3 \qquad\qquad (3-33)$$

Other laboratory model test results such as those conducted by Hueckel (1957) and Kostyukov (1967) compared reasonably well with the equivalent free surface method of analysis.

Example

3.2 _____

Redo Example Problem 3.2 using the theory of Ovesen and Stromann (Section 3.3.1). Assume $W = 0$.

Solution

Calculation of $P_{a(H)}$ and $P_{a(V)}$

From Equations (3-5), (3-6), and (3-12)

$$P_{a(H)} = \frac{1}{2}\gamma H^2 K_a \cos \phi$$

$$P_{a(V)} = \frac{1}{2}\gamma H^2 K_a \sin \phi$$

For $\phi = 32°$, $K_a \approx 0.28$ (Figure 3-9)

$$P_{a(H)} = (\frac{1}{2})(105)(4)^2(0.28)(\cos 32°) = 199.46 \text{ lb/ft}$$

$$P_{a(V)} = (\frac{1}{2})(105)(4)^2(0.28)(\sin 32°) = 111.28 \text{ lb/ft}$$

Calculation of K_{pH}

From Equation (3-10)

$$K_{pH}\tan \delta = \frac{P_{a(V)} + W}{\frac{1}{2}\gamma H^2}$$

Assume $W \approx 0$

$$K_{pH}\tan \delta = \frac{111.28}{(0.5)(105)(4)^2} = 0.132$$

Using Figure 3-8, for $\phi = 32°$ and $K_{pH}\tan \delta = 0.132$, the value of $K_{pH} \approx 3.4$.

Calculation of Q_u

Using Equation (3-19)

$$Q_u = B[\frac{1}{2}\gamma H^2 K_{pH} \quad P_{a(H)}]\left[\frac{C_{ov} + 1}{C_{ov} + \frac{H}{h}}\right]\left[F\left(\frac{\frac{H}{h} + 1}{\frac{B}{h}}\right) + 1\right]$$

Assume loose sand condition; so $C_{ov} = 14$ and $F = 0.26$. Thus

$$Q_u = (5)[(\frac{1}{2})(105)(4)^2(3.4) \quad 111.28)](\frac{14 + 1}{14 + 2})[0.26(\frac{2 + 1}{2.5}) + 1]$$

$$= (5)(2744.72)(0.9375(1.312) = \textbf{16,880 lb}$$

Example

3.3

Redo Example Problem 3.2 using Meyerhof's procedure described in Section 3.3.2.

Solution

From Equation (3-21)

$$Q_u = B(\frac{1}{2}\gamma H^2 K_b)\left[\frac{C_{ov} + 1}{C_{ov} + \frac{H}{h}}\right]\left[F\left(\frac{\frac{H}{h} + 1}{\frac{B}{h}}\right) + 1\right]$$

From Figure 3-15, $K_b \approx 2.95$

$$Q_u = (5)[(\frac{1}{2})(105)(4)^2(2.95)](\frac{14 + 1}{14 + 2})[0.26(\frac{2 + 1}{2.5}) + 1]$$

$$= (5)(2478)(0.9375)(1.312) = \mathbf{15,239.7\ lb}$$

Example

3.4

Redo Example Problem 3.2 using the procedure of Neely et al. (1973) outlined in Section 3.3.4. Use

(a) the equivalent free surface solution

(b) the surcharge method

Solution

Part a. From Equation (3-30)

$$Q_u = \gamma Bh^2 [M_{\gamma q(strip)}]S_f$$

Use $m = 0$. From Figure 3-19, for $\phi = 32°$ and $H/h = 2$, $M_{\gamma q} \approx 9.9$. Also for $B/h = 2.5$, $H/h = 2$ and $S_f \approx 1.1$ (from Figure 3-20). So

$$Q_u = [(105)(5)(2)^2](9.9)(1.1) = \mathbf{22,869\ lb}$$

Part b. Assuming $\delta = \phi/2$, Figure 3-18 gives $M_{\gamma q} \approx 7.2$. So

$$Q_u = [(105)(5)(2)^2](7.2)(1.1) = \mathbf{16,632\ lb}$$

Note: From the solutions given in Example Problems 3.1, 3.2, 3.3, and 3.4, it can be seen that Teng's method yields the smallest value of $Q_u \approx 13,060$ lb. The methods of Ovesen and Stromann, Meyerhof, and Neely et al. (surcharge method) give an average value of $Q_u \approx 16,000$ lb. However, the equivalent free surface method of Neely et al. results in the highest value of $Q_u \approx 23,000$ lb.

3.4   NATURE OF PASSIVE PRESSURE DISTRIBUTION IN FRONT OF A SHALLOW VERTICAL ANCHOR

It is now clear that the passive pressure developed in front of a vertical anchor when it is subjected to a horizontal force is the primary contributing factor to its holding capacity. It is thus of interest to know the nature of actual distribution of the passive pressure on the face of an anchor plate. Hanna, Das and Foriero (1988) measured the passive pressure distribution on the face of a *vertical strip anchor* along with the horizontal displacement, and these results are shown in Figure 3-23. The passive pressure was measured by attaching several transducers to the anchor. For these laboratory tests, the following parameters apply:

$\gamma = 15.6$ kN/m^3 $H/h = 4$

$\phi = 41.2°$ $h = 152.4$ mm

Hueckel, Kwasniewski and Baran (1965) also presented results of similar pressure distribution on a *square anchor plate* embedded in sand. Figure 3-24 shows one of their laboratory test results. The parameters for this test were as follows:

$\gamma = 16.38$ kN/m^3 $H/h = 2.5$

$\phi = 34.2°$ $B = h = 300$ mm

Horizontal displacement = 70 mm

From the above laboratory observations, the following conclusions can be drawn:

1.   The exact passive pressure distribution on the face of an anchor plate does not follow the classical pattern.

2.   The nature of the pressure distribution diagram will be approximately the same irrespective of the horizontal displacement.

3.5   DEEP VERTICAL ANCHOR

For a vertical anchor located at a shallow depth, the failure surface at ultimate load extends to the ground surface (Section 3.1). Under such conditions, the breakout factor F_q introduced in Section 3.3 increases with embedment ratio H/h (Figure 3-25). However at a given embedment ratio, $H/h = (H/h)_{cr}$, the magnitude of F_q remains practically constant. The maximum value

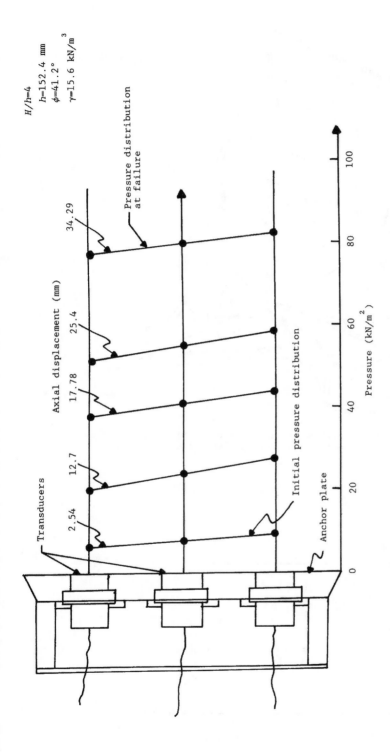

Figure 3-23 Nature of passive pressure distribution in front of a shallow vertical anchor (after Hanna, Das and Foriero, 1988)

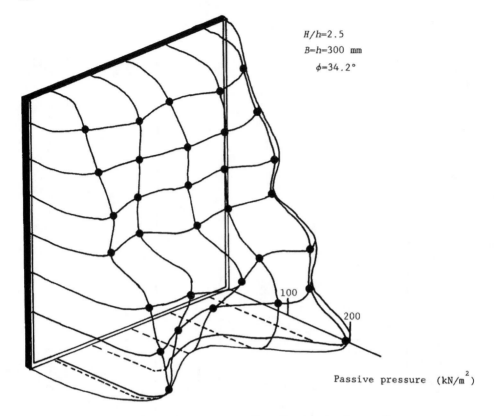

$H/h=2.5$
$B=h=300$ mm
$\phi=34.2°$

100

200

Passive pressure (kN/m^2)

Figure 3-24 Nature of passive pressure distribution in front of a shallow vertical anchor as observed by Hueckel et al. (1965)

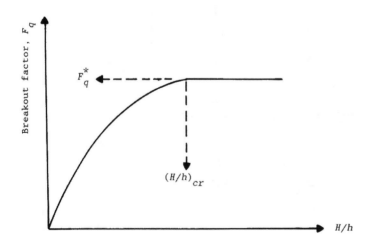

Figure 3-25 Nature of variation of F_q with H/h

of F_q may be denoted by F_q^*. The critical embedment ratio, $(H/h)_{cr}$, and the maximum breakout factor, F_q^*, are functions of ϕ and B/h. Anchors located at $H/h < (H/h)_{cr}$ are called *deep anchors* and, for this condition, at ultimate load local shear failure in soil around the anchor takes place.

Ovesen (1964) derived a relationship for the breakout factor of rectangular anchors as

$$F_q^* = \frac{Q_u}{\gamma(Bh)H} = S_f K_o e^{\pi \tan \phi} \cdot \tan^2(45 + \frac{\phi}{2}) \cdot d_c \tag{3-34}$$

where K_o = earth pressure coefficient at rest (which may be taken as $1 - \sin \phi$)

$$d_c = 1.6 + 4.1 \tan^4 \phi \tag{3-35}$$

$$S_f = \text{shape factor} \approx 1 + 0.2\frac{h}{B} \tag{3-36}$$

Ovesen (1964) however suggested that, for deep anchors, the shape factor S_f may be assumed to be one for all anchors, irrespective of the B/h ratio. Figure 3-26 shows the plot of F_q^* obtained from Equation (3-34) assuming $S_f = 1$.

Meyerhof (1973) extended his theory on shallow anchors (Section 3.3.2) to deep anchor condition and expressed the variation of $Q_u/\gamma(Bh)(H - h/2)$ as a function of the soil friction angle ϕ. However, for $H/h = 5$, the ratio of $Q_u/\gamma(Bh)(H - h/2)$ to $Q_u/\gamma(Bh)H$ is about 1.11 (Das, Seeley and Das, 1977). So, for all practical purposes

$$F_q^* = \frac{Q_u}{\gamma(Bh)(H - \frac{h}{2})} \approx \frac{Q_u}{\gamma(Bh)(H)}$$

Figure 3-27 shows the variation of F_q^* (for square and strip anchors with the soil friction angle ϕ).

Biarez et al. (1965) analyzed the characteristic rotational mechanism for *deep strip anchors* $(H/h > 7)$ as shown in Figure 3-28 by considering the couple necessary for the rotation of a soil cylinder. According to this solution (also see Dickin and Leung, 1985)

$$F_{q(strip)}^* = 4\pi(\frac{h}{H})(\frac{H}{h} - 1)\tan \phi \tag{3-37}$$

It appears that a shape factor similar to the type given by Equation (3-36) may be added to Equation (3-37) to obtain the breakout factor for rectangular anchors, or

112

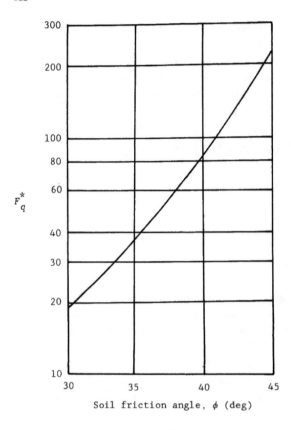

Soil friction angle, ϕ (deg)

Figure 3-26   Variation of F_q^* with soil friction angle [Equation (3-34)]

$$F_{q(rectangular)}^* = 4\pi(\frac{h}{H})(\frac{H}{h} \quad 1)\tan \phi(1 + 0.2\frac{h}{B})$$ (3-38)

A comparison of the breakout factors shown in Figures 3-26 and 3-27 shows that, for a given value of ϕ, $F_{q(strip)-Oyesen}^* > F_{q(strip)-Meyerhof}^*$. Das (1983) compiled the limited results of the F_q given by Das, Seeley and Das (1977) and Akinmusuru (1978), and this correlation is shown in Figure 3-29. Note that these results are for *square anchors only* ($B/h = 1$) and are based on small-scale laboratory model tests. These experimental results show that both of the theories predict a higher value of F_q^* than those obtained theoretically.

Das (1983) also gave a correlation for the critical embedment ratio of square anchors (based on model test results) in the form

$$(H/h)_{cr-square} = 5.5 + 0.166(\phi \quad 30) \quad (\text{for } \phi = 30\text{-}45°)$$ (3-39)

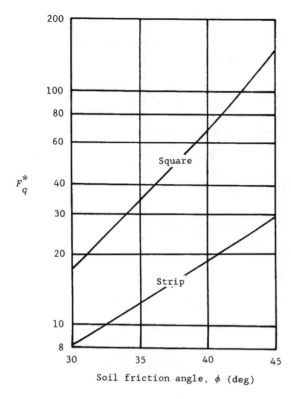

Figure 3-27 Meyerhof's values of F_q^* (1973)

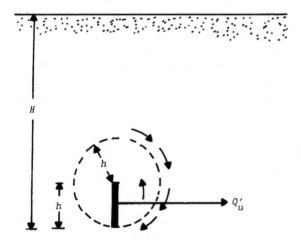

Figure 3-28 Failure mechanism around deep anchor as assumed by Biarez et al. (1965)

114

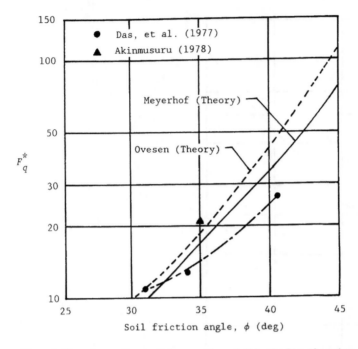

Figure 3-29   Comparison of Ovesen's and Meyerhof's theories with laboratory model test results for the variation of F_q^*

where ϕ is in degrees.   Based on the experiences of the author, it can be said that the critical embedment ratio of strip anchors is about 20-30% higher than that of square anchors under similar conditions.

3.6   LOAD-DISPLACEMENT RELATIONSHIP

In many instances, design restrictions allow limited horizontal movement of the anchors. Neely et al. (1973) showed the typical nature of variation of the load versus horizontal displacement diagrams from their laboratory model tests (Figure 3-30).   According to their results, three types of load-displacement diagrams may be observed for vertical anchors in sand.   They are:

1.   For anchors with $B/h < 2$ and $H/h < 2$, the load increases with displacement up to a maximum value (Q_u) and remains constant thereafter.

2.   For anchors with $B/h < 2$ and $H/h > 2$, the load increases with displacement up to a maximum value (Q_u), after which the load-displacement diagram becomes practically linear.

3.   For anchors with $B/h > 2$, at all values of H/h, the load increases with displacement to reach a peak value (Q_u) and decreases thereafter with displacement.

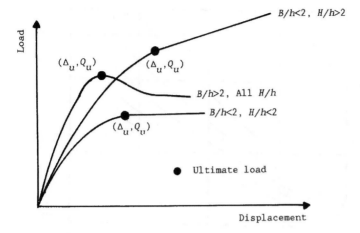

Figure 3-30 Typical nature of load versus displacement diagram for shallow anchors based on the observations of Neely et al. (1973)

The displacement of the anchor corresponding to the load Q_u may be referred to as Δ_u. The magnitudes of Δ_u obtained from the laboratory tests of Neely et al. (1973) are shown in a nondimensional form in Figure 3-31.

Based on their model test results, Das and Seeley (1975) recommended that, for $1 \leq B/h \leq 5$ and $1 \leq H/h \leq 5$, the load-displacement relationship can be expressed in the form

$$\frac{Q}{Q_u} = \frac{\dfrac{\Delta}{\Delta_u}}{0.15 + 0.85(\dfrac{\Delta}{\Delta_u})} \tag{3-40}$$

where Δ = displacement at load Q

Figure 3-32 shows a plot of Q/Q_u versus Δ/Δ_u based on Equation (3-40). It is important to realize that, in obtaining the preceding empirical relationship, some scattering of test results were observed and caution should be taken in using the equation.

Example

3.5

Refer to Example Problem 3.2.

(a) Determine the anchor displacement at ultimate load Q_u

(b) Determine the allowable load Q for an anchor displacement of 1 inch.

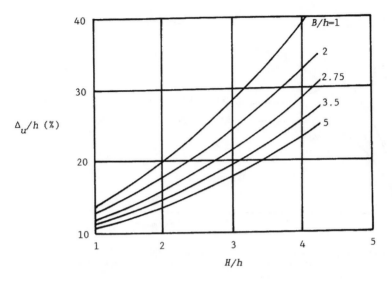

Figure 3-31   Nondimensional plot of Δ_u/h versus H/h for various values of B/h (after Neely et al., 1973)

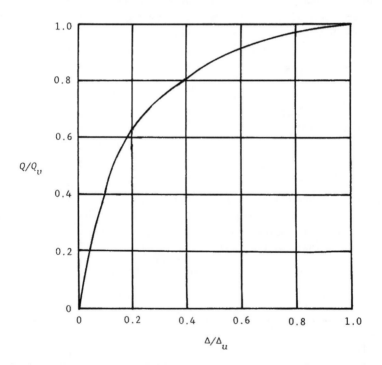

Figure 3-32   Plot of Q/Q_u versus Δ/Δ_u [Equation (3-40)]

Solution

Part a.   Referring to Figure 3-31, for $B/h = 2.5$ and $H/h = 2$, $\Delta_u/\Delta \approx 15\%$ = 0.15.   So

$$\Delta_u \approx 0.15(2 \times 12) = \textbf{3.6 in.}$$

Part b. $\Delta = 1$ in. $\Delta/\Delta_u = 1/3.6 = 0.278$

$$\frac{Q}{Q_u} = \frac{\dfrac{\Delta}{\Delta_u}}{0.15 + 0.85(\dfrac{\Delta}{\Delta_u})} = \frac{0.278}{0.15 + (0.85)(0.278)} = 0.72$$

$$Q = (0.72)(Q_u) = (0.72)(16,880) \approx \textbf{12,154 lb}$$

3.7   DESIGN CONSIDERATIONS

Angle of Friction

Prediction of the holding capacity of anchors for design needs careful consideration of the soil friction angle ϕ. Model test results generally overpredict the holding capacity when compared to the prototype. This is due primarily to the *scale effect*. Dickin and Leung (1983) conducted centrifuge model tests on vertical anchor plates, the results of which are shown in Figures 3-33 and 3-34. It is obvious that, as the anchor size (h) increases, the holding capacity decreases. This is true for all H/h and B/h ratios. The reason this type of behavior occurs is that, in the case of prototypes, the failure in the soil mass is *progressive*, and the applicable value of ϕ is not the peak value (that is, ϕ_{peak}). Furthermore, model tests are conducted at low stress levels. In reality, for soils, Mohr's failure envelope is actually curved as shown in Figure 3-35. This means that the peak friction angle, $\phi = \phi_{peak}$, obtained at a lower stress level is higher than that obtained at a higher stress level. From Figure 3-35, note that

$$\phi_{peak} = \phi_2 = tan^{-1}(\frac{\tau_2}{\sigma_2}) < \phi_{peak} = \phi_1 = tan^{-1}(\frac{\tau_1}{\sigma_1}) \quad (\text{since } \sigma_2 > \sigma_1)$$

Hence, for similar soil conditions, a lower peak value of ϕ may be expected in the case of a prototype when compared to that of a model. As an example, the variation of the plane strain peak friction angle for dense Erith sand (in which the test results shown in Figures 3-34 and 3-35 were conducted) with confining pressure is shown in Figure 3-36. Keeping the above considerations in mind, for continuous anchors Dickin and Leung (1985) suggested that the most appropriate friction angle to use for predicting the prototype anchor capacity is the mobilized plane-strain friction angle, ϕ_{mp}, or

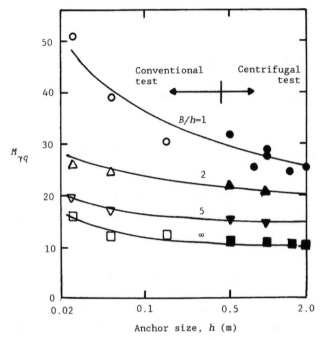

Figure 3-33 Variation of $M_{\gamma q}$ with anchor size; $H/h = 2$ (after Dickin and Leung, 1983)

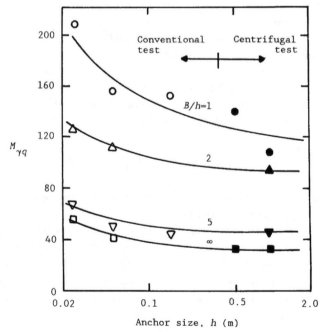

Figure 3-34 Variation of $M_{\gamma q}$ with anchor size; $H/h = 4$ (after Dickin and Leung, 1983)

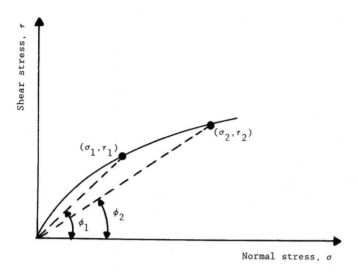

Figure 3-35 Curvilinear Mohr's failure envelope

$$\phi_{mp} = \phi_{ps}(1 \quad P_r) + P_r\phi_{cp} \tag{3-40}$$

where ϕ_{ps} = plane strain peak friction angle
ϕ_{cp} = critical state friction angle
P_r = progressivity index ≈ 0.8 (Rowe, 1969)

Figure 3-36 also shows the variation of ϕ_{cp} and ϕ_{mp} for Erith sand [based on Equation (3-40)].

Based on the study of Dickin and Leung (1985), the following general conclusions can be drawn:

1. For prototype strip anchors, all the theories (except for Biarez et al.) predict higher values of $M_{\gamma q}$ when the peak plane strain friction angle is used for calculation.

2. Neely et al.'s surcharge method (with $\delta = \phi/2$), Ovesen and Stromann's method, and Meyerhof's method give fairly good agreement with the experimental results when $\phi = \phi_{mp}$ is used for calculation. However, the theory of Neely et al. (based on the equivalent free surface method with $m = 1$) highly overestimates the experimental results.

Shape Factor

In most cases of construction, the soil will be compacted after placement of the anchor. In such cases, according to Ovesen and Stromann (1972), for single anchors [Equation (3-17)]

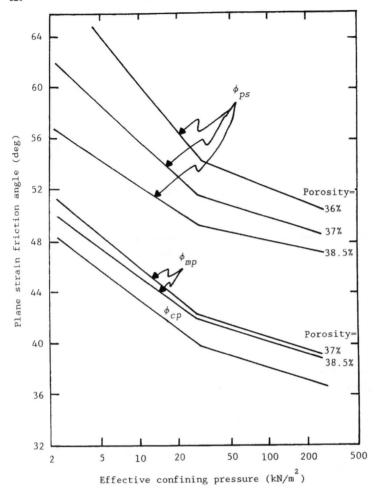

Figure 3-36 Variation of plane strain friction angle with effective confining pressure for Erith sand (after Dickin and Leung, 1985)

$$S_f = 0.42 \left[\frac{\frac{H}{h} + 1}{\frac{B}{h}} \right] + 1$$

Using the preceding equation, the variations of S_f with H/h and B/h for shallow anchors were calculated and are shown in Figure 3-37. For comparison purposes, the shape factors proposed by Neely et al. (1973) are also plotted in the figure. It appears that, for a given H/h and B/h, the shape factor given by Ovesen and Stromann is higher than that given by Neely et al. (1973). This is primarily due to the fact that Neely et al. assumed that the behavior of anchors with $B/h = 5$ is essentially the same as that of a strip anchor.

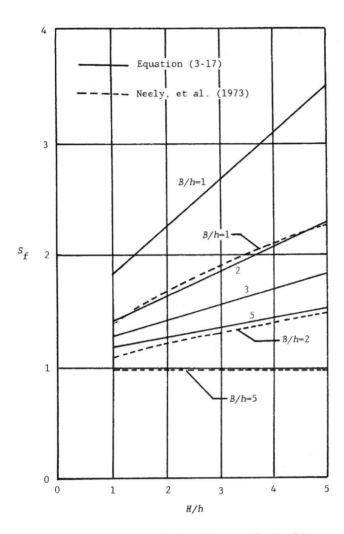

Figure 3-37   Comparison of shape factor relationships

General Recommendations

Although many factors affecting the ultimate holding capacity of anchors are yet to be considered and determined by further research, the following tentative design recommendations may be made based on the available information.

1. For routine works, plane strain tests on sand are rarely conducted in the laboratory to determine the friction angle. For that reason, it is recommended that the triaxial peak friction angle at a confining pressure of about 10 psi (100 kN/m^2) be determined. The

$\phi_{peak(triaxial)}$ will be about 10% less than the magnitude of $\phi_{peak(plane\ strain)}$.

2. The peak friction angle determined from triaxial tests may be used to determine the ultimate holding capacity for *single anchors* by using Ovesen and Stromann's procedure [Equation (3-19)]. For *strip anchors*, from Equation (3-19), note that

$$Q_u' = \frac{1}{2}\gamma H^2 (K_{pH} \quad K_a \cos \phi) \left[\frac{C_{ov} + 1}{C_{ov} + \frac{H}{h}}\right]$$

or

$$F_q = \frac{Q_u'}{\gamma h H} = 0.5(\frac{H}{h})(K_{pH} \quad K_a \cos \phi) \left[\frac{C_{ov} + 1}{C_{ov} + \frac{H}{h}}\right]$$

The maximum value of F_q should be equal to $F_{q(strip)}^*$ given by Meyerhof (1973) and shown in Figure 3-27, or

$$F_{q(strip)} = 0.5(\frac{H}{h})(K_{pH} \quad K_a \cos \phi) \left[\frac{C_{ov} + 1}{C_{ov} + \frac{H}{h}}\right] \le F_{q(strip)-Meyherhof}^*$$

In a similar manner

$$F_{q(square)} = \frac{Q_u}{\gamma h^2 H}$$

$$= 0.5(\frac{H}{h})(K_{pH} \quad K_a \cos \phi) \left[\frac{C_{ov} + 1}{C_{ov} + \frac{H}{h}}\right] [F(\frac{H}{h} + 1) + 1]$$

$$\le F_{q(square)-Meyerhof}^*$$

For rectangular anchors, Meyerhof's values of F_q^* should be interpolated and used as the upper limit, or

$$F_{q(rectangular)} = \frac{Q_u}{\gamma h B h}$$

$$= 0.5(\frac{H}{h})(K_{pH} \quad K_a \cos \phi) \left[\frac{C_{ov} + 1}{C_{ov} + \frac{H}{h}}\right] \left[F\left[\frac{\frac{H}{h} + 1}{\frac{B}{h}}\right] + 1\right]$$

$$\le F_{q(rectangular-Meyerhof)}^*$$

The surcharge method with $\delta = \phi/2$, or the equivalent free surface method with $m = 0$ as provided by Neely et al. (1972) along with their recommended shape factors, may also be used for obtaining the ultimate holding capacity of anchors at $H/h \leq 3$, or

$$Q_u = \gamma Bh^2 M_{\gamma q(strip)} \cdot S_f$$

3. The ultimate holding capacity for single anchors determined from Step 2 needs to be reduced to account for scale effects. Neely et al. (1973) suggested that a 31% reduction of Q_u for ten-fold increase in size should be considered as a probable upper limit of the magnitude of the scale effects. Dickin and Leung (1985) suggested that the reduction due to scale effects could still be larger. It appears than a reduction of about 30% in Q_u in conjunction with the triaxial peak friction angle may be appropriate. So

$$Q_{u(field)} \approx 0.3 Q_{u(\text{Step 2})}$$

4. For allowable load, a factor of safety (F_s) of about 2 may be used, or

$$Q_{all(field)} \approx \frac{0.3 Q_{u(\text{Step 2})}}{2} = 0.15 Q_{u(\text{Step 2})}$$

According to Equation (3-40)

$$\frac{Q}{Q_u} = \frac{\dfrac{\Delta}{\Delta_u}}{0.15 + 0.85(\dfrac{\Delta}{\Delta_u})}$$

With $Q/Q_u = 1/2$

$$\frac{\Delta}{\Delta_u} = 0.075 + 0.425(\frac{\Delta}{\Delta_u})$$

$$\frac{\Delta}{\Delta_u} = \frac{0.075}{0.575} = 0.13$$

or

$$\Delta = 0.13 \Delta_u$$

A deflection of $\Delta = 0.13\Delta_u$ will roughly correspond to a value of $\Delta \leq 0.065h$ for square anchors and $\Delta \leq 0.03h$ for strip anchors.

124

5. When shallow anchors are used in groups, the procedure of Ovesen and Stromann outlined in Section 3.2.1 should be used to obtain Q_u by using the triaxial peak friction angle (Step 1). A reduction factor of about 30% for scale effects (Step 3) and a factor of safety of about 2 (Step 4) should be used to obtain Q_{all} in the field.

3.8 EFFECT OF ANCHOR INCLINATION

Limited studies are available relating to the holding capacity of inclined plate anchors subjected to horizontal pull. Hueckel (1957) conducted laboratory tests with model anchor plates having dimensions of 150 mm × 150 mm ($B/h = 1$). The average embedment depth H' was at $1.5h$. The pullout tests were conducted with $\theta = 0°$, ±30° and ±45°. Figure 3-38 shows the positive and negative orientations of the anchor inclination with respect to the vertical. Figure 3-39a shows the nature of variation of $Q_{u(\theta)}/Q_{u(\theta=0°)}$ with θ. Das and Seeley (1975) also conducted similar tests with model anchor plates measuring 51 mm × 51 mm ($B/h = 1$), 51 mm × 153 mm ($B/h = 3$) and 51 mm × 255 mm ($B/h = 5$) with $\theta = 0°$, ±15° and ±30°, and $H'/h = 0.5$, 1.5, 2.5 and 3.5. The variations of $Q_{u(\theta)}/Q_{u(\theta=0°)}$ from these tests are shown in Figures 3-39b, 3-39c and 3-39d. Although theoretical developments to quantify these results are not yet available, the following general conclusions can be drawn (based on Figure 3-39).

1. For a given anchor plate, H'/h, and soil compaction, negatively inclined anchors offer more resistance to horizontal pull than positively inclined anchors.

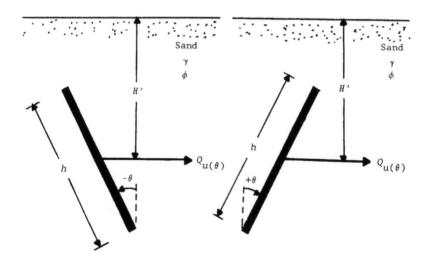

Figure 3-38 Positively- and negatively-oriented anchor

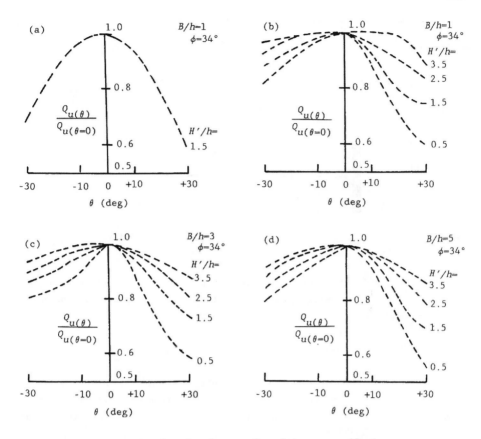

Figure 3-39 Nondimensionalized form of ultimate pullout resistance for inclined anchors (*Note*: (a) From Hueckel, 1957; (b), (c) and (d) from Das and Seeley, 1975)

2. For given values of θ and B/h, the ratio of $Q_{u(\theta)}/Q_{u(\theta=0°)}$ increases with the increase of H'/h.

3. For given values of θ and H'/h, the value of $Q_{u(\theta)}/Q_{u(\theta=0°)}$ increases with the decrease of the width-to-height ratio of the anchor plate.

ANCHORS IN CLAY (UNDRAINED COHESION; $\phi = 0$)

3.9 ULTIMATE HOLDING CAPACITY

Figure 3-40 shows the geometric parameters of a vertical plate anchor embedded in saturated clay. The undrained shear strength of the clay is c_u ($\phi = 0$

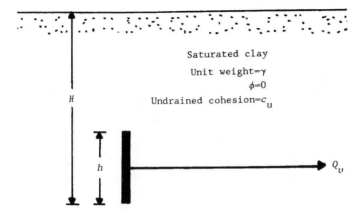

Figure 3-40   Geometric parameters of vertical anchor embedded in saturated clay

condition). The height, width, and depth of embedment of the plate anchor are h, B, and H, respectively. Let the ultimate holding capacity of the anchor plate be equal to Q_u, or

$$Q_u' = \frac{Q}{B}$$
(3-41)

where Q_u' = ultimate holding capacity per unit width at right angles to the cross section shown in Figure 3-40. The ultimate holding capacity for an anchor embedded in clay can be expressed in a nondimensional form (Tschebotarioff, 1973) as

$$F_c = \frac{Q_u'}{c_u h}$$
(3-42)

where F_c = breakout factor

Large-scale field test results to determine the ultimate holding capacity of plate anchors in undrained clay are scarce. Some of the initial laboratory model test results on this subject were reported by Mackenzie (1955). The tests were conducted on strip anchors (plane strain condition) in two different clayey soils. The average plot of these laboratory model test results is given in Figure 3-41. From this plot it may be seen that the breakout factor F_c increases with the embedment ratio (H/h) up to a maximum limit ($F_c = F_c^*$) and remains constant thereafter. Thus, as in the case of vertical anchors in sand, the failure mode in soil can be divided into two categories: (i) shallow anchor condition and (ii) deep anchor condition. The dividing line between the two modes of failure is the critical embedment ratio, $(H/h)_{cr}$. For $H/h \leq (H/h)_{cr}$, the anchor behaves as a shallow anchor, and the

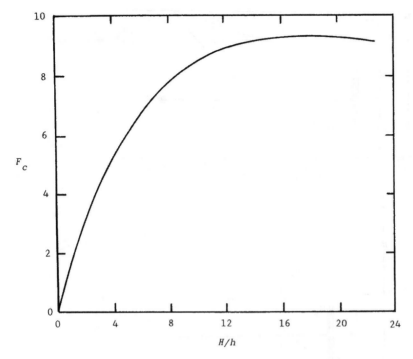

Figure 3-41   Average plot of F_c versus H/h for strip anchors in clay (ϕ = 0) based on Mackenzie (1955) and Tschebotarioff (1973)

failure in soil surrounding the anchor takes place at ultimate load. These failure modes are shown in Figure 3-42. For Mackenzie's model tests, the magnitudes of F_c^* and $(H/h)_{cr}$ were approximately 9 and 12, respectively.

Based on a limited number of laboratory model tests, Meyerhof (1973) proposed the following conservative estimate of the breakout factor and critical embedment ratio for square and strip anchors.

For square anchor

$$
\left.
\begin{aligned}
F_c &= 1.2(\tfrac{H}{h}) \le 9 \\[2mm]
(\tfrac{H}{h})_{cr} &= 7.5
\end{aligned}
\right\}
\qquad (3\text{-}43)
$$

128

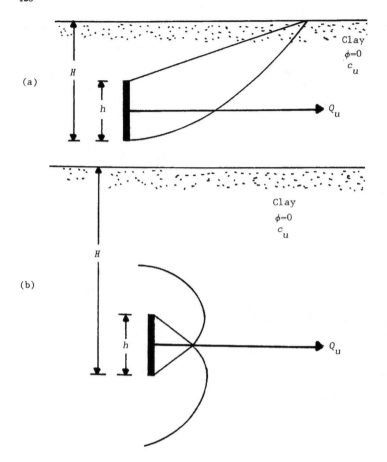

Figure 3-42   Failure modes: (a) shallow anchor;   (b) deep anchor

For strip anchor

$$F_c = 1.0\left(\frac{H}{h}\right) \le 8 \left.\vphantom{\frac{H}{h}}\right\}$$
$$\left(\frac{H}{h}\right)_{cr} = 8 \qquad\qquad$$

$$(3\text{-}44)$$

The findings on the holding capacity of anchors discussed in the preceding paragraphs leave some unanswered questions. They are

1.   What is the dependence of $(H/h)_{cr}$ on the undrained shear strength of clay, and also the width-to-height ratio of the plate?
2.   What is the nature of variation of F_c for rectangular anchors?

For that reason, Das, Tarquin and Moreno (1985) conducted a number of small-scale laboratory model tests, the results of which provide considerable insight to the problem. Figure 3-43 shows the variation of the breakout factor F_c for square anchors with the undrained shear strength of the clay (Das, Tarquin and Moreno, 1985). From this figure it can be seen that the critical embedment ratio in soft and medium clays increases with c_u up to a maximum limit and remains constant thereafter. This general behavior can be expressed as

$$(\tfrac{H}{h})_{cr\text{-}S} = 4.7 + 2.9 \times 19^{-3} c_u \le 7 \tag{3-45}$$

where $(H/h)_{cr\text{-}S}$ = critical embedment ratio of square anchors (that is, $h = B$) and c_u is in lb/ft^2 In SI units, the preceding expression can be stated as

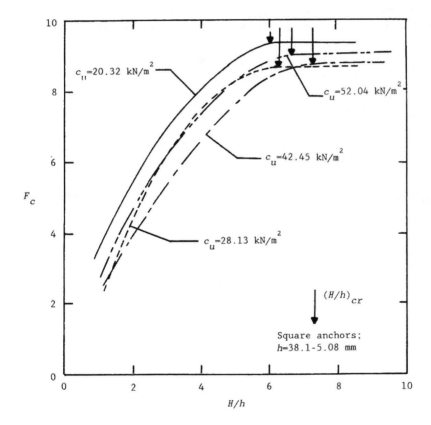

Figure 3-43   Model test results of Das et al. (1985)--variation of F_c versus H/h for square anchor

$$(\frac{H}{h})_{cr-S} = 4.7 + 0.0606c_u \leq 7 \tag{3-46}$$

where c_u is in kN/m^2

The upper limit of $(H/h)_{cr-S} = 7$ is generally consistent with the recommendations of Meyerhof (1973) as given by Equation (3-43). Based on limited model test results, Das et al. (1985) also showed that, for similar undrained shear strength of clay, the critical embedment ratio of rectangular anchors and that of square anchors can be approximated as

$$\frac{(\frac{H}{h})_{cr-R}}{(\frac{H}{h})_{cr-S}} = [0.9 + 0.1(\frac{B}{h})] \leq 1.31 \tag{3-47}$$

From Equations (3-46) and (3-47) it is obvious that, for strip anchors with $B/h = \infty$, $(H/h)_{cr-strip} = 1.31(H/h)_{cr-S}$. In medium stiff to stiff clays, $(H/h)_{cr-S}$ is about 7, so $(H/h)_{cr-strip} \approx 9.17$. This value of 9.17 falls between Meyerhof's recommended value of 8 [Equation (3-44)] and the value of 12 obtained from Mackenzie's tests.

Das et al. (1985) recommended that the breakout factor for deep rectangular anchors can be given by the relationship

$$F_{c(R)}^* = F_{c(S)}^*[0.825 + 0.175(\frac{h}{H})] \tag{3-48}$$

where $F_{c(R)}^*$ = breakout factor for deep rectangular anchor
$F_{c(S)}^*$ = breakout factor for deep square anchor = 9

For shallow square and rectangular anchors, the breakout factor can be expressed by two nondimensional parameters:

$$\beta = \frac{F_c}{F_c^*} \tag{3-49}$$

and

$$\alpha = \frac{(\frac{H}{h})}{(\frac{H}{h})_{cr}} \tag{3-50}$$

The relationships between α and β can be expressed as (Das et al., 1985)

$$\beta = \frac{\alpha}{0.41 + 0.59\alpha} \tag{3-51}$$

Or

$$F_c = \frac{F_c^*(\frac{H}{h})}{0.41(\frac{H}{h})_{cr} + 0.59(\frac{H}{h})} \qquad (3\text{-}52)$$

3.10   STEP-BY-STEP PROCEDURE FOR ESTIMATION OF ULTIMATE LOAD

With limited experimental results available at the present time, the following step-by-step procedure may be used for estimation of the ultimate holding capacity of single rectangular anchors in clay ($\phi = 0$ condition).

1.   Estimate the embedment ratio (H/h) and the width-to-height ratio (B/h).

2.   Estimate the undrained shear strength of clay (c_u).

3.   Using Equation (3.45) [or Equation (3.46) and c_u in Step 2], determine the critical embedment ratio of a square anchor.

4.   With known values of B/h (Step 1) and Equation (3-47), obtain the ratio of $(H/h)_{cr\text{-}R}/(H/h)_{cr\text{-}S}$.

5.   With known values of $(H/h)_{cr\text{-}S}$ from Step 3 and the ratio of $(H/h)_{cr\text{-}R}/(H/h)_{cr\text{-}S}$ from Step 4, calculate $(H/h)_{cr\text{-}R}$.

6.   If the actual embedment ratio H/h (Step 1) is equal to or greater than the $(H/h)_{cr\text{-}R}$ calculated in Step 5, it is a deep anchor condition. So

$$F_{c(R)}^* = 9.0[0.825 + 0.175(\frac{h}{B})]$$

or

$$Q_u = 9c_u Bh[0.825 + 0.175(\frac{h}{B})]$$

7.   If the actual H/h is less than the critical embedment ratio calculated in Step 5, it is a shallow anchor. Equation (3-52) can be used to estimate Q_u, or

$$Q_u = c_u Bh\left[\frac{F_{c(R)}^*(\frac{H}{h})}{0.41(\frac{H}{h})_{cr\text{-}R} + 0.59(\frac{H}{h})}\right]$$

where $F_{c(R)}^* = 9[0.825 + 0.175(\frac{h}{B})]$

Example

3.6

Estimate the ultimate breakout load of a rectangular anchor plate with the following: $H = 1.2$ m, $h = 0.3$ m, $B = 0.6$ m, and $c_u = 48$ kN/m^2

Solution

$$\frac{H}{h} = \frac{1.2 \text{ m}}{0.3 \text{ m}} = 4$$

$$\frac{B}{h} = \frac{0.6 \text{ m}}{0.3 \text{ m}} = 2$$

From Equation (3-46)

$$(\frac{H}{h})_{cr-S} = 4.7 + 0.0606c_u = 4.7 + 0.0606(48) = 7.609 > 7$$

So, use $(\frac{H}{b})_{cr-S} = 7$

From Equation (3-47)

$$\frac{(\frac{H}{h})_{cr-R}}{(\frac{H}{h})_{cr-S}} = [0.9 + 0.1(\frac{B}{h})] = 0.9 + (0.1)(2) = 1.1$$

Hence

$$(\frac{H}{h})_{cr-R} = (1.1)(7) = 7.7$$

The actual H/h is 2, so this is a shallow anchor.

$$Q_u = c_u Bh \left[\frac{F^*_{c(R)}(\frac{H}{h})}{0.41(\frac{H}{h})_{cr-R} + 0.59(\frac{H}{h})} \right]$$

$$F^*_{c(R)} = 9[0.825 + 0.175(\frac{b}{H})] = 9[0.825)(0.175(\frac{0.3}{0.6})] = 8.21$$

So

$$Q_u = [(48)(0.6)(0.3)][\frac{(8.21)(4)}{(0.41)(7.7) + (0.59)(4)}] = \textbf{51.43 kN}$$

3.11 LIMITATIONS OF THE EXISTING STUDY

There are certain limitations and uncertainties in applying the existing results in literature to the estimation of the allowable holding capacity of a plate anchor embedded in clay (under undrained conditions), and they are:

1.   Since most of the relationships cited in Section 3.9 are based on small-scale model test results, the scale effect has not yet been investigated. However such effects in clay soils are expected to be minimal.

2. All of the model test results thus far reported are based on tests on single anchors. However vertical plate anchors may, and are, used in groups. Figure 3-11 shows the plan view of a group of vertical anchor plates subjected to horizontal pull. The failure surfaces in soil around the anchor at ultimate load may overlap each other. In effect, this will reduce the magnitude of Q_u. Thus

$$Q_{u(actual)} = \eta Q_{u(isolated)} \qquad (3\text{-}53)$$

where η = efficiency factor = $f(\frac{S}{B}, \frac{H}{h})$

The efficiency factor η is Equation (3-53) has not as yet been investigated. For that reason a conservative estimate of $F_{c(R)}$ would be to assume $h/B = 0$ (strip case) and, thus, from Equation (3-48)

$$F^*_{c(R)} \approx 7.43$$

3. A factor of safety of at least 3 should be used to determine the allowable holding capacity.

REFERENCES

Akinmusuru, J.O., 1978. Horizontally loaded vertical plate anchors in sand. *J. Geotech. Engrg. Div.*, ASCE, 104(2):283-286.

Biarez, I., Boucraut, L.M., and Negre, R., 1965. Limiting equilibrium of vertical barriers subjected to translation and rotation forces. *Proc.*, VI Intl. Conf. Soil Mech. Found. Engrg., Montreal, Canada, 2:386-372.

Caquot, A. and Kerisel, L., 1949. *Traite de macanique des sols*. Gauthier-Villars, Paris, France.

Das, B.M., 1983. Holding capacity of vertical anchor slabs in granular soil. *Proc.*, Coastal Structures '83, ASCE:379-392.

Das, B.M., 1975. Pullout resistance of vertical anchors. *J. Geotech. Eng.*, ASCE, 101(1):87-91.

Das, B.M. and Seeley, G.R., 1975. Load-displacement relationship for vertical anchor plates. *J. Geotech. Engrg. Div.*, ASCE, 101(7):711-715.

Das, B.M., Seeley, G.R., and Das, S.C., 1977 Ultimate resistance of deep vertical anchor in sand. *Soils and Found.*, Japan, 17(2):52-56.

Das, B.M., Tarquin, A.J., and Moreno, R., 1985. Model tests for pullout resistance of vertical anchors in clay. *Civ. Engrg. for Pract. Design Engrs.*, Pergamon Press, New York, 4(2):191-209.

Dickin, E. and Leung, C.F., 1985. Evaluation of design methods for vertical anchor plates. *J. Geotech. Engrg.*, ASCE, 111(4):500-520.

134

Dickin, E. and Leung, C.F., 1983. Centrifugal model tests on vertical anchor plates. *J. Geotech. Engrg.*, ASCE, 109(12):1503-1525.

Hanna, A.M., Das, B.M., and Foriero, A., 1988. Behavior of shallow inclined plate anchors in sand. *Special Topics in Foundations*, Geotech. Spec. Tech. Pub. No. 16, ASCE:54-72.

Hueckel, S., 1957. Model tests on anchoring capacity of vertical and inclined plates. *Proc.*, IV Intl. Conf. Soil Mech. Found. Engrg., London, England, 2:203-206.

Hueckel, S., Kwasniewski, J., and Baran, L., 1965. Distribution of passive earth pressure on the surface of a square vertical plate embedded in soil. *Proc.*, VI Intl. Conf. Soil Mech. Found. Engrg., Montreal, Canada, 2:381-385.

Kostyukov, V.D., 1967. Distribution of the density of sand in the sliding wedge in front of anchor plates. *Soil Mech. Found. Engrg.* (in Russian) 1:12-13.

Mackenzie, T.R., 1955. Strength of deadman anchors in clay. M.S. thesis, Princeton University, USA.

Meyerhof, G.G., 1973. Uplift resistance of inclined anchors and piles. *Proc.*, VIII Intl. Conf. Soil Mech. Found. Engrg., Moscow, USSR, 2.1:167-172.

Meyerhof, G.G., 1951. The ultimate bearing capacity of foundations. *Geotechnique*, 2(4):301-332.

Neely, W.J., Stuart, J.G., and Graham, J., 1973. Failure loads of vertical anchor plates in sand. *J. Soil Mech. Found. Div.*, ASCE, 99(9):669-685.

Ovesen, N.K., 1964. Anchor slabs, calculation methods and model tests. *Bull. 16*, Danish Geotech. Inst., Copenhagen, Denmark.

Ovesen, N.K. and Stromann, H., 1972. Design methods for vertical anchor slabs in sand. *Proc.*, Specialty Conf. on Performance of Earth and Earth-Supported Structures, ASCE, 2.1:1481-1500.

Rowe, P.W., 1969. The relationship between the shear strength of sands in triaxial compression, plane strain and direct shear. *Geotechnique*, 19(1):75-86.

Rowe, P.W., 1952. Anchored sheet pile walls. *Proc.*, Institute of Civil Engineers, London, England, 1(1):27-70.

Sokolovskii, V.V., 1965. *Statics of granular media*. Pergamon Press, New York.

Teng, W.C., 1962. *Foundation design*. Prentice-Hall, Englewood Cliffs, New Jersey, USA.

Tschebotarioff, G.P., 1973. *Foundations, retaining structures and earth structures*. McGraw-Hill, New York, USA.

Chapter 4
INCLINED PLATE ANCHORS

4.1 INTRODUCTION

As pointed out in Chapter 1, in the construction of various types of foundations, plate anchors are sometimes placed at an inclination to the horizontal.
These anchors may be subjected to inclined or axial pull as shown in Figures
4-1a and 4-1b. However, in many cases, foundations most likely to be subjected to uplifting forces are constructed with horizontal and/or inclined
anchors with the assumption that the pullout force will be transmitted axially
to the anchors. This chapter is devoted primarily to a review and compilation
of existing theoretical and experimental results relating to the ultimate
holding capacity of *inclined plate anchors subjected to axial pull*.

For an inclined anchor subjected to axial pull, the gross ultimate holding capacity can be expressed as (Figure 4-1b)

$$Q_{u(g)} = Q_u + W_a \cos \psi$$

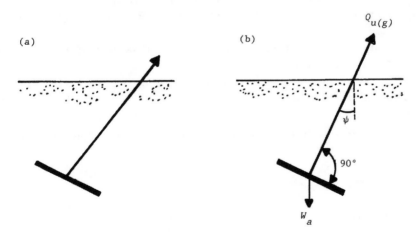

(a)   (b)

Figure 4-1   Inclined plate anchor subjected to (a) inclined pull;
(b) axial pull

where $Q_{u(g)}$ = gross ultimate holding capacity

Q_u = net ultimate holding capacity

W_a = self-weight of anchor

ANCHORS IN SAND

4.2   INCLINED ANCHORS--AXISYMMETRIC CASE (ANALYSIS OF HARVEY AND BURLEY)

Harvey and Burley (1973) made an analysis of the holding capacity of *shallow inclined circular anchor plates*, the details of which will be presented in this section. Figure 4-2 shows a vertical section through the assumed failure zone corresponding to the *major axis of the ground surface failure ellipse*. The diameter of the circular anchor is equal to h. The anchor is inclined at an angle ψ with respect to the horizontal, and the average depth of embedment is equal to H'. *AC* and *BD* are assumed failure surfaces which are arcs of circles. The arcs *AC* and *BD* make angles of 90° at *A* and *B* and intersect the ground surface at angles of 45-ϕ/2 (Rankine's passive state assumption). This assumption is similar to that of Balla's (1961) for horizontal anchors presented in Chapter 2. The assumption of the failure surface in soil can be further simplified by replacing the *curvilinear surface by a single curved surface* defined by the angle θ. So the trace of the simplified failure surface in Figure 4-2 can be given by the straight lines *AC* and *BD*.

In the analysis presented below, the following notations are used:

ξ = horizontal angle locating a typical sector of the failure zone measured from the major axis of surface failure ellipse

ω = angle of inclination of typical sector of failure zone relative to the vertical axis

η = angle of inclination of pullout axis to the plane perpendicular to the failure zone axis

α = angle between the curvilinear surface of sliding and the horizontal surface in the plane of failure sector

Now, considering a typical sector of the failure zone located at an angle ξ from the vertical section

$$\tan \omega = \tan \psi \cdot \sin \xi \tag{4-1}$$

$$\sin \eta = \sin \psi \cdot \cos \xi \tag{4-2}$$

$$\tan \alpha = \tan(45 \; \frac{\phi}{2}) \cdot \sqrt{(1 + \tan^2 \psi \cdot \sin^2 \xi)} \tag{4-3}$$

137

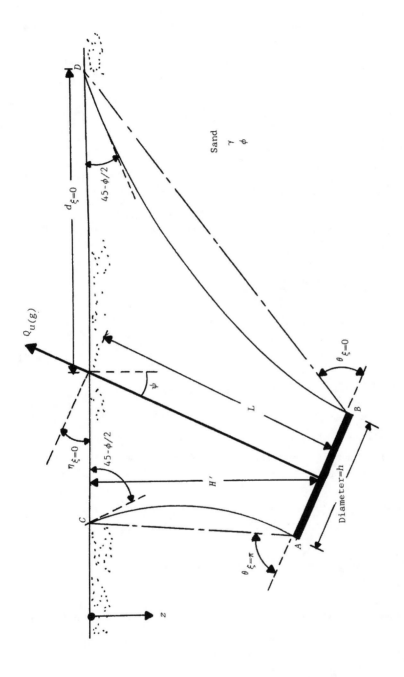

Figure 4-2 Inclined shallow circular plate anchor

138

and

$$\theta = \frac{\pi}{4} + \frac{\alpha}{2} + \frac{\eta}{2} \qquad (4-4)$$

The volume of a typical sector of the failure zone subtending an angle $d\xi$ can be given as

$$\delta V = \left[\frac{d^2}{6}(L + \frac{h}{2} \cdot \tan\,\theta)\cos\,\eta \quad \frac{h^3}{48}\,\tan\,\theta \cdot \sin\,\eta \right]\delta\xi \qquad (4-5)$$

where

$$d = \frac{L + \frac{h}{2} \cdot \tan\,\theta}{\cos\,\eta \cdot \tan\,\theta \quad \sin\,\eta} \qquad (4-6)$$

Similarly, the area of the simplified curved surface of a sector of the failure zone is

$$\delta A = \frac{1}{2}(d + \frac{h}{2} \cdot \sec\,\eta)\left(\frac{d \cdot \cos\,\eta \quad \frac{h}{2}}{\cos\,\theta} \right)\delta\xi \qquad (4-7)$$

Similarly, the area of the sector of the anchor plate is

$$\delta A_a = \frac{1}{8}h^2 \cdot \delta\xi \qquad (4-8)$$

The lateral at-rest earth pressure at depth z is

$$p_z = K_o\gamma z \qquad (4-9)$$

where K_o = coefficient of at-rest earth pressure ($\approx 1 - \sin\,\phi$)
γ = unit weight of soil

So, the circumferential force on the sector due to earth pressure forces is equal to

$$\left[\frac{K_o\gamma d}{(L + \frac{h}{2} \cdot \tan\,\theta)} \right] \times \int_{z=0}^{z=L\cos\,\psi} (z \cdot \frac{h}{2} \cdot \tan\,\theta + zL \quad \frac{z^2}{\cos\,\psi})dz$$

From the preceding relationship, the radial force on the section can be determined as

$$\left[\frac{K_o\gamma d\,L^2\cos^2\psi}{2(L + \frac{h}{2} \cdot \tan\,\theta)} \right] [(\frac{h}{2})\tan\,\theta + \frac{L}{3}]d\xi$$

Now, the radial force which includes the effect of the weight component is

$$
F_R = \left\{ \left[\frac{K_o \gamma d \ L^2 \cos^2 \psi}{2(L + \frac{h}{2} \cdot \tan \theta)} \right] [(\frac{h}{2}) \tan \theta + \frac{L}{3}] \right.
$$

$$
+ \frac{1}{2}(\gamma)(\delta V) \sin \omega \cdot \cos \omega + \frac{1}{8} \ \gamma \cdot d \cdot \sin^2 \omega \cdot \cos \omega \cdot \cos^2 \psi \cdot (L + \frac{h}{2} \cdot \tan \theta)^2
$$

$$
\left. \frac{1}{8} \left[\frac{\gamma \cdot h^2 \cdot d \cdot \sin^2 \omega \cdot \cos \omega \cdot \cos^2 \psi \cdot \tan^2 \theta}{(L + \frac{h}{8} \cdot \tan \theta)} \right] [(\frac{h}{2}) \tan \theta + L] \right\} \qquad (4\text{-}10)
$$

If the anchor plate reaction on the adjacent soil is Q acting at an angle ϕ then

$$
Q \sin \theta = \gamma(\delta V) \cos \omega \cdot \sin(\phi + \theta \quad \eta) \quad F_R \cdot \cos(\theta + \phi \quad \eta)
$$

The net ultimate holding capacity can now be given as

$$
Q_u = \sum_{\zeta=0}^{\zeta=2\pi} Q \cos \phi \qquad (4\text{-}11)
$$

Equation (4-11) can be solved by using a computer program. The size of each sector can be defined by assigning a certain value to ξ. Note that L is equal to $H/\cos \psi$.

Harvey and Burley (1973) compared the analysis proposed above with the experimental results of Kanayan (1968) as well those of Baker and Kondner (1966). However, this procedure for determining the ultimate holding capacity of circular inclined anchor plates is rarely used in practice now.

4.3 MEYERHOF'S PROCEDURE

Figure 4-3 shows an inclined shallow strip anchor with a height h embedded in a c-ϕ soil. The bottom of the anchor plate is at a depth H measured from the ground surface. The average depth of embedment of the anchor is H'. The anchor is inclined at an angle ψ with respect to the horizontal and is subjected to an axial pullout force. For shallow anchor condition, the net ultimate holding capacity per unit width Q_u' at right angles to the cross section shown is (Meyerhof, 1973)

$$
Q_u' = P_p \quad P_a = cK_c H + \frac{1}{2}K_b \gamma H^2 + W \cos \psi \qquad (4\text{-}12)
$$

140

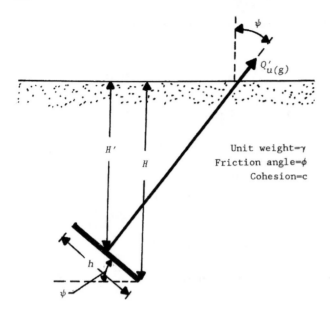

Figure 4-3   Inclined shallow strip anchor plate

where c = cohesion

γ = unit weight of soil

K_c, K_b = net earth pressure coefficients

W = weight of soil located directed above the anchor = $\gamma h H \cos \psi$

So

$$Q'_u = cK_c H + \frac{1}{2}K_b \gamma H^2 + \gamma h H \cos^2 \psi \qquad (4\text{-}13)$$

For granular soils, $c = 0$. So

$$Q'_u = \frac{1}{2}K_b \gamma H^2 + \gamma h H \cos^2 \psi \qquad (4\text{-}14)$$

The variations of K_b for shallow strip anchors can be obtained from the earth pressure coefficients for an inclined wall (Caquot and Kerisel, 1949; Sokolovskii, 1965). These values are shown in Figure 4-4 for $\psi = 20°$, $45°$, $75°$ and $90°$ Note that the variation of K_b for $\psi = 90°$ given in Figure 4-4 is the same as shown in Figure 3-15.

Equation (4-14) can be rewritten in the form

$$Q'_u = \frac{1}{2}K_b \gamma (H' + \frac{h \sin \psi}{2})^2 + \gamma h (H' + \frac{h \sin \psi}{2}) \cos^2 \psi \qquad (4\text{-}15)$$

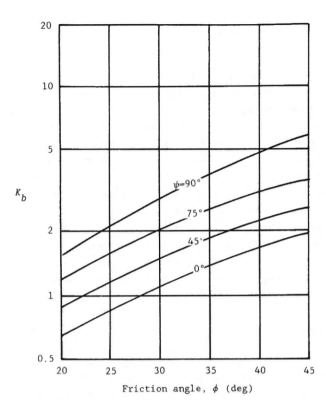

Figure 4-4 Variation of Meyerhof's earth pressure coefficient K_b

$$F'_q = \frac{Q'_u}{\gamma h H'} = \frac{1}{2}K_b(\frac{H'}{h})(1 + \frac{h}{2H'}\sin\psi)^2 + (1 + \frac{h}{2H'}\sin\psi)\cos^2\psi \qquad (4\text{-}16)$$

where F'_q = average breakout factor

However note that, with $\psi = 0$, $H = H'$ So

$$F'_q = F_q \qquad (\text{see Chapter 2}) \qquad (4\text{-}17)$$

Again, with $\psi = 90°$, $H' = H$ $h/2$. So

$$F'_q = \frac{Q'_u}{\gamma h(H \quad \frac{h}{2})} = \frac{Q'_u}{\gamma h H(1 \quad \frac{h}{2H})} = \frac{F_q}{1 \quad \frac{h}{2H}} \qquad (4\text{-}18)$$

where F_q = breakout factor as defined in Chapter 3

For $H/h \geq 5$

$$\frac{F'_q}{F_q} \leq 1.1$$

Hence, for shallow anchors, the average breakout factor can be calculated knowing the values of K_b, H'/h and ψ (for a given soil friction angle). As in the case of horizontal and vertical anchors (Chapters 2 and 3), for a given anchor orientation (that is, ψ) there exists a critical average embedment ratio $H'/h = (H'/h)_{cr}$ beyond which the average breakout factor will remain practically constant signifying *deep anchor behavior* by which local shear failure in soil takes place (Figure 4-5). For $H'/h \geq (H'/h)_{cr}$ and given values of ψ, the average breakout factor will remain practically constant ($F'_q = F'^{*}_q$). So, if the critical embedment ratio $(H'/h)_{cr}$ can be substituted into Equation (4-16) the average breakout factor for deep anchors can be estimated.

Based on the experimental observations of Meyerhof and Adams (1968), with $\psi = 0$, the critical embedment ratio $[(H'/h)_{cr} = (H/h)_{cr}]$ for square anchors in loose sand is about 4 and increases to about 8 in dense sand (as discussed in Chapter 2). However, for strip anchors (for $\psi = 0$)

$$(\frac{H'}{h})_{cr-strip} \approx 1.5(\frac{H'}{h})_{cr-square}$$

For anchor inclination $\psi > 0$, the magnitude of $(H'/h)_{cr}$ gradually decreases. Using the preceding conditions, the variations of the magnitude of F'^{*}_q for $\psi = 0°$, 45° and 90° for *deep strip anchors* have been determined and are shown in Figure 4-6.

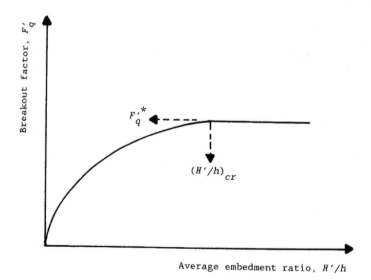

Average embedment ratio, H'/h

Figure 4-5   Nature of variation of F'_q with H'/h

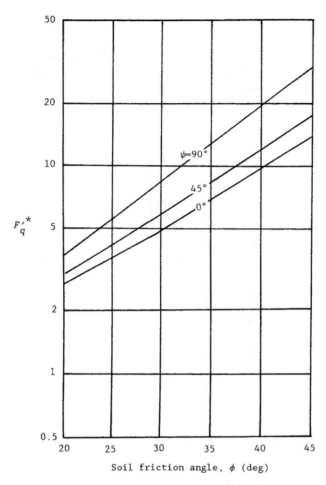

Figure 4-6   Variation of Meyerhof's $F_q{'}^{*}$ with soil friction angle ϕ for strip anchor

Again, for a given value of ψ

$$F_{q(square)}^{'*} = \frac{Q_u}{\gamma A H'} = F_{q(strip)}^{'*} \cdot S_f \tag{4-19}$$

where A = area of the anchor plate

S_f = shape factor

For horizontal anchors ($\psi = 0°$) the shape factor increases roughly with H'/h (Meyerhof and Adams, 1968) up to the above-mentioned critical depths. A similar relationship which is applicable for establishing the shape factors of square vertical anchors ($\psi = 90°$) can be developed from the work of Brinch Hansen (1961) on horizontally loaded rigid piles. Using the above-stated

144

shape factors and the critical embedment ratios (H'/h) in Equation (4-16), the variation of $F_q'^*$ for *deep square anchors* with $\psi = 0°$ and $90°$ have been calculated and are shown in Figure 4.7. The magnitude of $F_q'^*$ for a rectangular anchor slab having a width B and height h can be interpolated between the values of strips and squares in proportion to the B/h ratio.

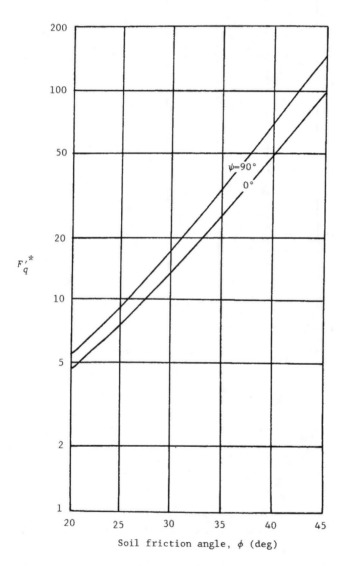

Figure 4-7   Variation of Meyerhof's $F_q'^*$ with soil friction angle ϕ for square anchor

Example

4.1

A strip anchor is shown in Figure 4-8. Given: $\phi = 35°$, $\gamma = 17$ kN/m^3, $h = 0.4$ m, $H' = 1$ m. Determine the variation of the net ultimate load Q'_u for $\psi = 0°$, $45°$, $75°$ and $90°$

Solution

From Equation (4-15)

$$Q'_u = \frac{1}{2}K_b\gamma(H' + \frac{h \sin \psi}{2})^2 + \gamma h(H' + \frac{h \sin \psi}{2})\cos^2 \psi$$

Referring to Figure 4-4, the variation of K_b with the anchor inclination ψ can be determined. Thus

ϕ (deg)	Anchor inclination, ψ (deg)	K_b
35	0	≈ 1.4
35	45	≈ 1.8
35	75	≈ 2.7
35	90	≈ 3.9

Using the above values of K_b, the magnitude of Q'_u can be determined.

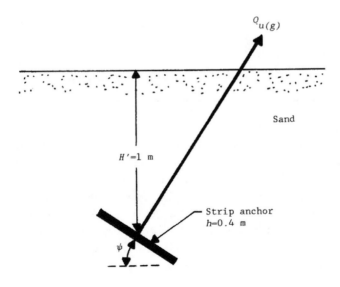

Figure 4-8

$\underline{\psi = 0}$

$$Q_u' = (\frac{1}{2})(1.4)(17)[1 + \frac{(0.4)(\sin 0)}{2}]^2$$

$$+ (17)(0.4)[1 + \frac{(0.4)(\sin 0)}{2}]\cos^2 0$$

$$= 11.9 + 6.8 = \textbf{18.7 kN/m}$$

$\underline{\psi = 45°}$

$$Q_u' = (\frac{1}{2})(1.8)(17)[1 + \frac{(0.4)(\sin 45)}{2}]^2$$

$$+ (17)(0.4)[1 + \frac{(0.4)(\sin 45)}{2}]\cos^2 45$$

$$= 19.93 + 3.88 = \textbf{23.81 kN/m}$$

$\underline{\psi = 75°}$

$$Q_u' = (\frac{1}{2})(2.7)(17)[1 + \frac{(0.4)(\sin 75)}{2}]^2$$

$$+ (17)(0.4)[1 + \frac{(0.4)(\sin 75)}{2}]\cos^2 75$$

$$= 32.67 + 0.54 = \textbf{33.21 kN/m}$$

$\underline{\psi = 90°}$

$$Q_u' = (\frac{1}{2})(3.9)(17)[1 + \frac{(0.4)(\sin 90)}{2}]^2$$

$$+ (17)(0.4)[1 + \frac{(0.4)(\sin 90)}{2}]\cos^2 90 = \textbf{47.74 kN/m}$$

4.4   ANALYSIS OF HANNA ET AL.

Hanna, Das and Foriero (1988) developed an analytical method for estimation of the ultimate holding capacity of shallow *inclined strip anchors* with ψ varying from zero to 60° In order to explain this method of analysis, let us consider a shallow strip anchor (Figure 4-9). At ultimate load the actual failure surface in soil will be somewhat similar to ab' and cd' However, along planes ab and cd, the passive forces per unit width of the anchor will be P_1 and P_2, respectively. These resultant forces will be inclined at an angle δ to the normal drawn to ab and cd. So, it can be written that

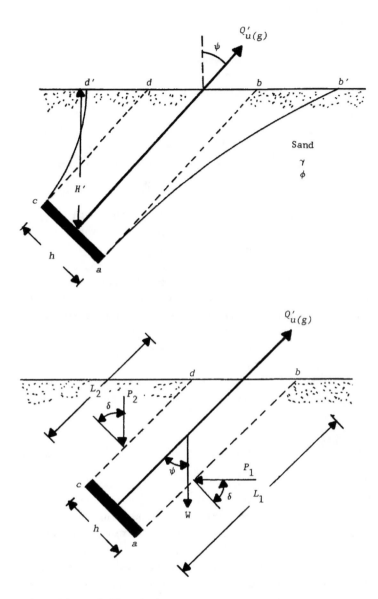

Figure 4-9   Shallow inclined strip anchor

$$Q'_u = P_1 \sin \delta + P_2 \sin \delta + W \cos \psi \qquad (4\text{-}20)$$

where Q'_u = net ultimate holding capacity per unit width

W = weight of the soil in zone *abcd* per unit width of the anchor at right angles to the cross section shown

148

Note that

$$W = \frac{1}{2}(L_1 + L_2)h\gamma \cos \psi \tag{4-21}$$

$$P_1 = \frac{1}{2}R_\gamma K_p \gamma L_1^2 \tag{4-22}$$

$$P_2 = \frac{1}{2}R_\gamma K_p \gamma L_2^2 \tag{4-23}$$

where K_p = passive earth pressure coefficient with $\delta = \phi$ = soil friction angle
R_γ = reduction factor for K_p which is a function of δ/ϕ

The magnitudes of K_p and R_γ can be determined from the earth pressure tables of Caquot and Kerisel (1949). It needs to be pointed out that, if the earth pressure analysis is conducted based on the actual failure surfaces ab' and cd' (Figure 4-9), then the mobilized friction angle δ would be equal to ϕ. On the other hand, if the analysis is made on the assumed failure surfaces ab and cd, then the mobilized angle of friction δ cited in Equation (4-20) is the average value. The locally mobilized friction angle δ_Z would be similar to the type shown in Figure 4-10. At points b and d, $\delta_Z = \lambda\phi$ (where λ is a function which depends on ψ); and, at points a and c, $\delta = \phi$ (since points a and c are located on the actual failure surface).

In the study of Hanna et al. (1988), the average value of δ was determined by combining some laboratory model test results in conjunction with Equations (4-20), (4-22) and (4-23). Or

$$R_\gamma K_p \sin \delta = \frac{Q_u' \quad 0.5(L_1 + L_2)h\cdot\gamma \cos \psi}{0.5\gamma(L_1^2 + L_2^2)} \tag{4-24}$$

The right-hand side of Equation (4-24) was obtained from the laboratory model test results, and the left-hand side was determined by assuming several δ/ϕ values and the passive earth pressure coefficient tables of Caquot and Kerisel (1949). The solution proceeded by trial and error until both sides of Equation (4-24) were equal. Once the average value of δ was determined from Equation (4-24), the variations of the locally mobilized angle of shearing resistance δ_Z were determined in the following manner.

$$P_1 + P_2 = \frac{1}{2}R_\gamma K_p \gamma(L_1^2 + L_2^2) = \gamma\left[\int_0^{L_1} K_{p(Z)}Z\cdot dZ + \int_0^{L_2} K_{p(Z)}Z\cdot dZ \right] \tag{4-25}$$

The magnitude of the term $\frac{1}{2}R_\gamma K_p \gamma(L_1^2 + L_2^2)$ shown in of Equation (4-25) was obtained by knowing the average value of δ. Use of laboratory experimental

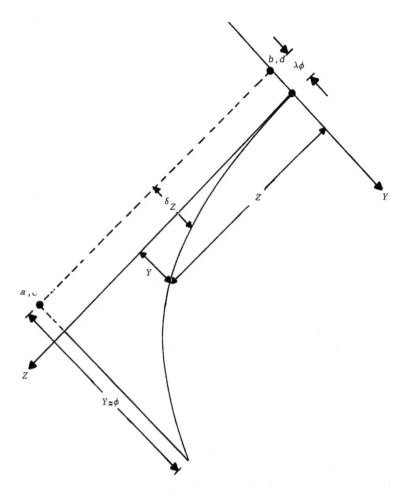

Figure 4-10   Nature of distribution of δ_Z

results and several trials and errors showed that the solution to Equation (4-25) can be found if

$$\lambda = (\frac{\psi°}{90})^3 + e^{-5\tan \phi} \tag{4-26}$$

and

$$Y = \frac{A'Z}{1 \quad B'Z} \tag{4-27}$$

where $A' = \frac{\lambda\phi}{\beta}$ \hspace{1cm} (4-28)

and

$$B' = \frac{1}{\beta} \tag{4-29}$$

The term β is a constant which is determined by boundary conditions, or along ab (Figure 4-9)

$$\beta = \frac{L_1}{1 - \lambda} \tag{4-30}$$

and along cd (Figure 4-9)

$$\beta = \frac{L_2}{1 - \lambda} \tag{4-31}$$

The purpose of the above exercise was to predict δ and, hence, K_p and R_γ for soil friction angles other than those used in the model tests of Hanna et al. (1988). Following is a step-by-step procedure for determining these values:

1. Assume a value of ϕ.
2. From Equation (4-26) calculate λ.
3. Calculate the variation of Y from Equations (4-27), (4-28), (4-29), (4-30) and (4-31).
4. Calculate δ_Z as

$$\delta_Z = \lambda\phi + Y \tag{4-32}$$

5. With the value of δ_Z obtained from Equation (4-32), obtain the magnitude of $K_{p(Z)}$ from the tables of Caquot and Kerisel (1949).
6. Using a computer program, calculate $R_\gamma K_p$ from Equation (4-25), or

$$R_\gamma K_p = \frac{\displaystyle\int_0^{L_1} K_{p(Z)} Z \cdot dZ + \int_0^{L_2} K_{p(Z)} Z \cdot dZ}{0.5(L_1^2 + L_2^2)} \tag{4-33}$$

7. Once the right-hand side of Equation (4-33) is known, determine the average δ using the passive earth pressure tables of Caquot and Kerisel (1949).

The results of this type of analysis, if conducted, will be as shown in Figure 4-11, which is a plot of δ/ϕ versus ψ for various soil friction angles ϕ. The analysis can be further simplified if we assume that

$$K_s \sin\phi = R_\gamma K_p \sin\delta \tag{4-34}$$

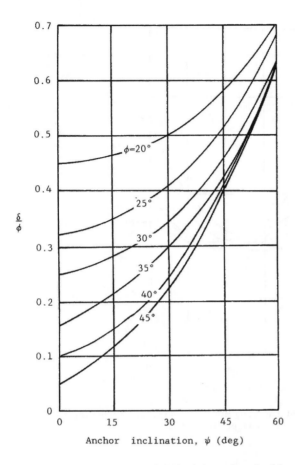

Figure 4-11 Variation of δ/ϕ with anchor inclination ψ

Or

$$K_s = \frac{R_\gamma K_p \sin \delta}{\sin \phi} \tag{4-35}$$

where K_s = punching uplift coefficient

The variations of K_s thus obtained for various values of ψ and ϕ are shown in Figure 4-12. Now, combining Equations (4-20), (4-21), (4-22), (4-23) and (4-34), we obtain

$$Q'_u = \frac{1}{2}\gamma K_s \sin \phi(L_1^2 + L_2^2) + \frac{1}{2}\gamma(L_1 + L_2)h \cdot \cos \psi \tag{4-36}$$

In Equation (4-36), note that (Figure 4-13)

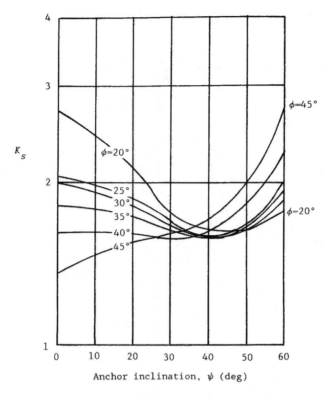

Figure 4-12   Variation of punching uplift coefficient according to the theory of Hanna et al. (1988)

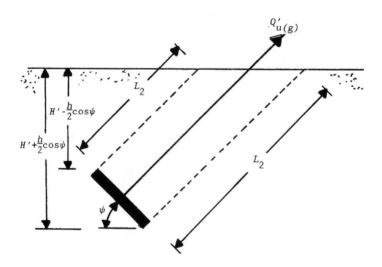

Figure 4-13   Definition of L_1 and L_2 [Equations (4-37) and (4-38)]

$$L_1 = \frac{H' + \frac{h}{2} \sin \psi}{\cos \psi} \qquad (4\text{-}37)$$

$$L_2 = \frac{H' \quad \frac{h}{2} \sin \psi}{\cos \psi} \qquad (4\text{-}38)$$

So

$$Q'_u = \frac{1}{2}\gamma K_s \sin \phi \left[\left(\frac{H' + \frac{h}{2} \sin \psi}{\cos \psi} \right)^2 + \left(\frac{H' \quad \frac{h}{2} \sin \psi}{\cos \psi} \right)^2 \right] + \frac{1}{2}\gamma \left[\left(\frac{H' + \frac{h}{2} \sin \psi}{\cos \psi} \right) \right.$$

$$\left. + \left(\frac{H' \quad \frac{h}{2} \sin \psi}{\cos \psi} \right) \right] h \cdot \cos \psi$$

or

$$Q'_u = \gamma K_s \frac{\sin \phi}{\cos^2 \psi} \left(H'^2 + \frac{h^2}{4} \sin^2 \psi \right) + \gamma H'h \qquad (4\text{-}39)$$

Example
4.2

With the parameters for sand and anchor given in Example Problem 4.1, determine Q'_u for $\psi = 0°$, $45°$ and $60°$ using the theory of Hanna et al.

Solution

Given: $H' = 1$ m; $h = 0.4$; $\gamma = 17$ kN/m^3; $\phi = 35°$ From Figure 4-12 for $\phi = 35°$, the variations of K_s are as follows:

Anchor inclination, ψ (deg)	K_s
0	1.8
45	1.6
60	2.0

Now, from Equation (4-39)

$$Q'_u = \gamma K_s \frac{\sin \phi}{\cos^2 \psi} \left(H' + \frac{h^2}{4} \sin^2 \psi \right) + \gamma H'h$$

For $\psi = 0°$

$$Q_u' = (17)(1.8)\left[\frac{\sin 35}{\cos^2 0}\right]\left[(1)^2 + \frac{(0.4)^2}{4}\sin^2 0\right] + (17)(1)(0.4)$$

$$= 17.55 + 6.8 = \mathbf{24.35 \ kN/m}$$

For $\psi = 45°$

$$Q_u' = (17)(1.6)\left[\frac{\sin 35}{\cos^2 45}\right]\left[(1)^2 + \frac{(0.4)^2}{4}(\sin 45)^2\right] + (17)(1)(0.4)$$

$$= 31.83 + 6.8 = \mathbf{38.63 \ kN/m}$$

For $\psi = 60°$

$$Q_u' = (17)(2)\left[\frac{\sin 35}{\cos^2 60}\right]\left[(1)^2 + \frac{(0.4)^2}{4}(\sin 60)^2\right] + (17)(1)(0.4)$$

$$= \mathbf{87.63 \ kN/m}$$

4.5 OTHER EMPIRICAL RELATIONSHIPS

A simple empirical relationship for estimating the ultimate holding capacity of shallow inclined anchors embedded in sand was proposed by Maiah, Das and Picornell (1986), which is of the form

$$Q_{u-\psi} = Q_{u-\psi=0°} + [Q_{u-\psi=90°} \quad Q_{u-\psi=0°}](\frac{\psi°}{90})^2$$

$$\text{(for a given value of } H'/h) \tag{4-40}$$

where $Q_{u-\psi}$ = net ultimate holding capacity of anchor inclination of ψ with respect to the horizontal (Figure 4-14a)

$Q_{u-\psi=0°}$ = net ultimate uplift capacity of horizontal anchor (that is, $\psi = 0°$; see Figure 4-14b)

$Q_{u-\psi=90°}$ = net ultimate holding capacity of vertical anchor (that is, $\psi=90°$; see Figure 4-14c)

The above relationship was originally developed for shallow strip anchors; however, the author feels that it can also be applied to rectangular anchors.

In order to predict $Q_{u-\psi=0°}$, the relationship presented by Meyerhof and Adams (1968) given in Chapter 2 can be used, or

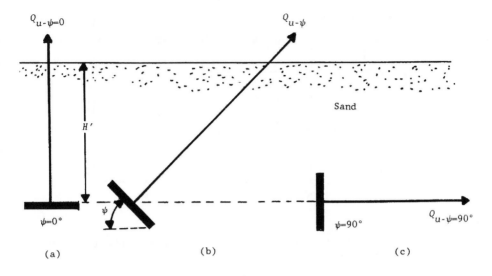

Figure 4-14 Definition of $Q_{u-\psi}$, $Q_{u-\psi=0°}$ and $Q_{u-\psi=90°}$

$$Q_{u-\psi=0°} = \gamma H'^2 [2(1 + m \frac{H'}{h})h + B \quad h]K_u \tan \phi + hBH'\gamma$$

(for rectangular anchors)   (4-41)

$$Q'_{u-\psi=0°} = \gamma H'^2 K_u \tan \phi + hH'\gamma \quad \text{(for strip anchors)} \tag{4-42}$$

where B = length of anchor plate (dimension at right angle to the cross sec-
        tion shown in Figure 4-14)

   h = width of anchor plate

   m = a coefficient for obtaining the shape factor (see Table 4-1)

   K_u = uplift coefficient (see Figure 4-15)

Table 4-1   Variation of m with
            soil friction angle

Soil friction angle, ϕ (deg)	m
20	0.05
25	0.1
30	0.15
35	0.25
40	0.35
45	0.50
48	0.6

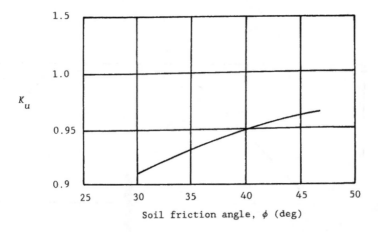

Figure 4-15   Meyerhof's uplift coefficient K_u for horizontal ($\psi=0°$) plate anchor

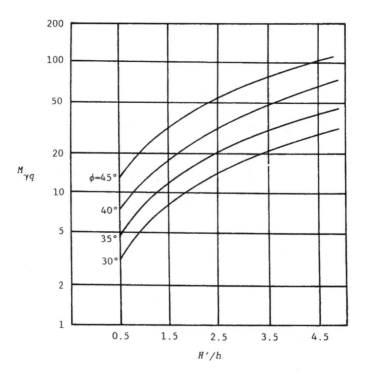

Figure 4-16   Variation of Neely et al.'s force coefficient with H'/h ($\delta = \phi$)

In order to calculate the magnitude of $Q_{u-\psi=90°}$, the theory of Neely, Stuart and Graham (1973) can be used. This theory is based on the surcharge method with $\delta = \phi$ as discussed in Chapter 3, or

$$Q_{u-\psi=90°} = \gamma h^2 B M_{\gamma q} S_f \qquad (4\text{-}43)$$

where $M_{\gamma q}$ = force coefficient (Figure 4-16)
S_f = shape factor (Figure 4-17)

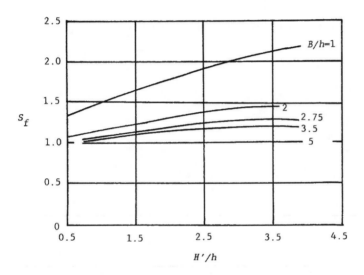

Figure 4-17   Variation of S_f with H'/h and B/h (based on the study of Neely et al., 1973)

Example

4.3

Solve Example Problem 4.1 using Equations (4-40), (4-42) and (4-43).

Solution

From Equation (4-42)

$$Q'_{u-\psi=0°} = \gamma H'^2 K_u \tan \phi + h H' \gamma$$

For $\phi = 35°$, $K_u = 0.93$ (Figure 4-15). $h = 0.4$ m, $H' = 1$ m, $\gamma = 17$ kN/m^3

158

So

$$Q'_{u-\psi=0°} = (17)(1)^2(0.93)(\tan 35) + (0.4)(1)(17) = 11.07 + 6.8$$
$$= 17.68 \text{ kN/m}$$

Again, from Figure 4-17, for $H'/h = 2.5$, $M_{\gamma q} \approx 20$. $S_f = 1$ (strip anchor) So, from Equation (4-43)

$$Q'_{u-\psi=90°} = \gamma h^2 B M_{\gamma q} S_f = (17)(0.4)^2 (1)(20)(1) = 54.4 \text{ kN/m}$$

Now we can use Equation (4-40) to estimate the variation of $Q'_{u-\psi}$.

ψ (deg)	$Q'_{u-\psi}$ (kN/m)
0	**17.68**
45	**26.86**
60	**34.00**
75	**43.13**
90	**54.4**

Example 4.4

Compare the results of Example Problems 4.1, 4.2 and 4.3.

Solution

The variation of $Q'_{u-\psi}$ with anchor inclination is shown in Figure 4-18. From the plot, the following conclusions can be drawn:

1. Meyerhof's theory for shallow inclined anchors [Equation (4-15)] and the theory of Maiah et al. [Equation (4-40)], in conjunction with Equations (4-42) and (4-43), yield fairly close results.
2. The theory of Hanna et al. [Equation (4-39)] provides excessively high values of $Q'_{u-\psi}$.

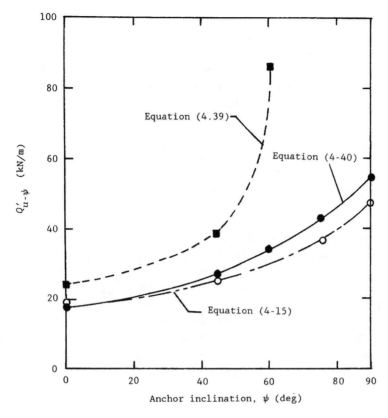

Figure 4-18

4.6 GENERAL REMARKS

As per the discussions provided in Sections 4.2 through 4.5, the following
general observations may be made.

1. The appropriate average critical embedment ratio $(H'/h)_{cr}$ for square
   and circular anchors with $\psi = 0°$ and $\psi = 90°$ are as follows:

   $\underline{\psi = 0°}$:   Loose sand   4

                  Dense sand   .   8

   $\underline{\psi = 90°}$:   Loose sand   4

                  Dense sand   6

2. In a similar manner, the appropriate values of $(H'/h)_{cr}$ for strip
   anchors are as follows:

   $\underline{\psi = 0°}$:   Loose sand   6

                  Dense sand   11 to 12

   $\underline{\psi = 90°}$:   Loose sand   4

                  Dense sand   8

3. For shallow strip anchors, the magnitude of $Q'_{u-\psi}$ can be estimated by using Equation (4-15) (Meyerhof's theory) or Equation (4-40) in conjunction with Equations (4-42) and (4-43).

4. For shallow rectangular anchors, the magnitude of $Q_{u-\psi}$ can be estimated by using Equations (4-40), (4-41) and (4-43).

5. For deep anchors, a similar relationship as in Equation (4-40) can be used, or

$$F'^*_{q-\psi} = F'^*_{q-\psi=0°} + (F'^*_{q-\psi=90°} \quad F'^*_{q-\psi=0°})(\frac{\psi°}{90})^2 \tag{4-44}$$

where $F'^*_{q-\psi}, F'^*_{q-\psi=0°}, F'^*_{q-\psi=90°}$ = breakout factor for deep anchors with inclinations of $\psi°$, 0° and 90° with respect to the horizontal

The preceding relation should be applicable for strip, square and rectangular anchors. For strip anchors $F'_{q-\psi=0°} = Q'_{u-\psi=0°}/\gamma hH'$ and $F'_{q-\psi=90°} = Q'_{u-\psi=90°}/\gamma hH'$ can be obtained from Figure 4-6. Similarly, for square anchors the magnitudes of $F'^*_{q-\psi=0°}$ and $F'^*_{q-\psi=90°}$ can be obtained from Figure 4-7. For rectangular anchors, interpolations need to be made to estimate the breakout factors for $\psi = 0°$ and 90°. Once $F'^*_{q-\psi}$ is determined, the magnitudes of the ultimate load can be obtained as:

Strip anchor
$$Q_{u-\psi} = F'^*_{q-\psi}\gamma hH' \tag{4-45}$$

Square anchor
$$Q_{u-\psi} = F'^*_{q-\psi}\gamma h^2 H' \tag{4-46}$$

Rectangular anchor
$$Q'_{u-\psi} = F^*_{q-\psi}\gamma hBH' \tag{4-47}$$

6. The anchor displacement Δ_u along the direction of the pull at ultimate load gradually increases with the anchor inclination ψ. Appropriate values of Δ_u/h for shallow anchor condition are as follows:

Anchor type	$\frac{\Delta_u}{h}$ at $\psi = 0°$	$\frac{\Delta_u}{h}$ at $\psi = 90°$
Strip	6% to 8%	10% to 25%
Square	8% to 10%	15% to 30%

The preceding approximate values of Δ_u/h are based on laboratory model tests conducted by the author. The magnitude of Δ_u/h increases with the increase of H'/h.

7   The present theories on inclined anchors are primarily based on observations made during model tests in the laboratory. Further full-scale tests are essential to verify the assumptions and results.

8.   Due to the uncertainties involved, a factor of safety of at least 3 may be used to obtain the allowable holding capacity.

9.   At the present time, experimental results are not available to consider the effects of center-to-center spacing of inclined anchors when they are placed in a row or a group.

INCLINED PLATE ANCHORS IN CLAY (ϕ = 0 CONDITION)

4.7   ULTIMATE HOLDING CAPACITY

Unlike cases where horizontal plate anchors are subjected to vertical uplift (Chapter 2) and vertical plate anchors are subjected to horizontal pull (Chapter 3), the existing studies relating to the holding capacity of inclined anchors embedded in clay and subjected to axial pull are fairly limited. Among them, the study by Das (1985) is fairly comprehensive and will be presented in this section. The results were based primarily on laboratory observations on square anchors embedded in saturated and near-saturated clay soils. According to the suggested procedure of Das (1985), the net ultimate holding capacity of an inclined rectangular anchor plate can be given as (Figure 4-19)

$$Q_u = Ac_u F'_c + W \cos \psi \qquad (4\text{-}48)$$

where A = area of the anchor plate = Bh

B = width of the anchor plate

c_u = undrained cohesion of the clay soil (ϕ = 0 condition)

F'_c = average breakout factor

W = weight of soil located immediately above the anchor

ψ = anchor inclination with respect to the horizontal

However

$$W = A\gamma H' \cos \psi \qquad (4\text{-}49)$$

where H' = average depth of embedment

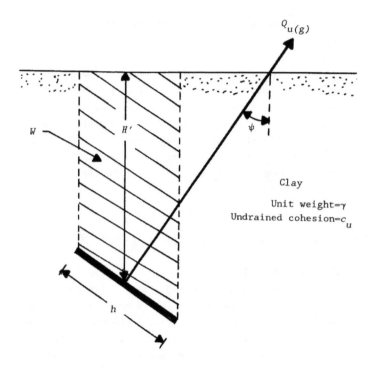

Figure 4-19   Inclined plate anchor in clay

Substituting Equation (4-49) into Equation (4-48), we obtain

$$F'_c = \frac{\dfrac{Q_u}{A} \quad \gamma H' \cos^2 \psi}{c_u}$$

For square anchors

$$F'_c = \frac{\dfrac{Q_u}{h^2} \quad \gamma H' \cos^2 \psi}{c_u} \tag{4-50}$$

For rectangular anchors

$$F'_c = \frac{\dfrac{Q_u}{Bh} \quad \gamma H' \cos^2 \psi}{c_u} \tag{4-51}$$

Similarly, for strip anchors

$$F'_C = \frac{\dfrac{Q'_u}{h} \quad \gamma H' \cos^2 \psi}{c_u} \tag{4-52}$$

The variation of the average breakout factor can be given as

$$F'_{c-\psi} = F'_{c-\psi=0°} + (F'_{c-\psi=90°} \quad F'_{c-\psi=0°})(\frac{\psi°}{90})^2 \tag{4-53}$$

The breakout factor $F'_{c-\psi}$ increases with the increase of the average embedment ratio H'/h to a maximum value $(F'^{*}_{c-\psi})$ at $(H'/h)_{cr}$ and remains constant thereafter (Figure 4-20).

The empirical procedures for estimating $F'_{c-\psi=0°}$ and $F'_{c-\psi=90°}$ were given in Chapters 2 and 3 [based on the studies of Das (1980) and Das et al. (1985)] and are summarized below.

Estimation of $F'_{c-\psi=0°}$

1. Calculate the critical average embedment ratio $(H'/h)_{cr-R}$ for a rectangular anchor $(h \times B)$

$$(\frac{H'}{h})_{cr-R} = (\frac{H'}{h})_{cr-S}[0.73 + 0.27(\frac{B}{h})] \leq 1.55(\frac{H'}{h})_{cr-S} \tag{4-54}$$

where

$$(\frac{H'}{h})_{cr-S} = 0.107 c_u + 2.5 \leq 7 \tag{4-55}$$

where c_u is in kN/m^2

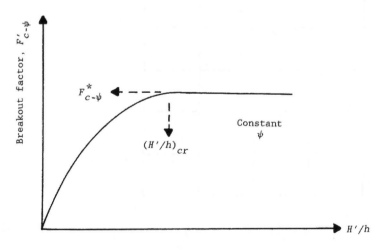

Figure 4-20 Variation of $F'_{c-\psi}$ with H'/h

2. If the actual H'/h is greater than $(H'/h)_{cr}$, then it is a deep anchor and

$$F'_{c-\psi=0°} = F'^{*}_{c-\psi=0°} = 7.56 + 1.44\left(\frac{h}{B}\right) \qquad (4\text{-}56)$$

If the actual H'/h is less than or equal to $(H'/h)_{cr}$, then it is a shallow anchor and

$$F'_{c-\psi=0°} = \left[7.56 + 1.44\left(\frac{h}{B}\right)\right]\beta \qquad (4\text{-}57)$$

where

$$\beta = f\left[\frac{\dfrac{H'}{h}}{\left(\dfrac{H'}{h}\right)_{cr}}\right] \qquad \text{(see Figure 4.21)}$$

Estimation of $F'_{c-\psi=90°}$

1. Calculate $(H'/h)_{cr-R}$ as

$$\frac{\left(\dfrac{H'}{h}\right)_{cr-R} + 0.5}{\left(\dfrac{H'}{h}\right)_{cr-S} + 0.5} = \left[0.9 + 0.1\left(\frac{B}{h}\right)\right] \leq 1.31 \qquad (4\text{-}58)$$

where $(H'/h)_{cr-S}$ = critical embedment ratio of a square anchor measuring $h \times h$

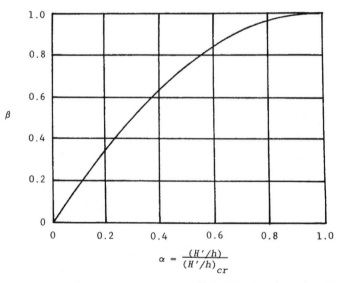

$$\alpha = \frac{(H'/h)}{(H'/h)_{cr}}$$

Figure 4-21   Variation of β with $(H'/h)/(H'/h)_{cr}$ for $\psi = 0°$

$$\left(\frac{H'}{h}\right)_{cr-S} = 4.2 + 0.0606c_u \leq 6.5 \tag{4-59}$$

where c_u is in kN/m^2

2. If the actual H'/h is greater than $(H'/h)_{cr}$, then it is a deep anchor.

$$F'_{c-\psi=90°} = (9)[0.825 + 0.175(\frac{h}{B})] \tag{4-60}$$

If the actual H'/h is less than or equal to $(H'/h)_{cr}$, then it is a shallow anchor. In that case

$$\frac{F'_{c-\psi=90°}}{F'^{*}_{c-\psi=90°}} = \frac{n'}{0.41 + 0.59n'} \tag{4-61}$$

where

$$n' = \frac{(\frac{H'}{h}) + 0.5}{(\frac{H'}{h})_{cr} + 0.5} \tag{4-62}$$

Once $F'_{c-\psi}$ is determined by using Equation (4-53), the magnitude of the ultimate holding capacity can be determined from Equations (4-50), (4-51) or (4-52).

Example

4.5

For an anchor embedded in a saturated clay, given, for the anchor: $h = 0.4$ m, $H' = 1.2$, $B = 0.8$ m, $\psi = 30°$ Given, for the clay: $c_u = 28$ kN/m^2, $\gamma = 18.4$ kN/m^3 Calculate the net ultimate holding capacity.

Solution

Calculation of $F'_{c-\psi=0°}$

From Equation (4-55)

$$\left(\frac{H'}{h}\right)_{cr-S} = 0.107c_u + 2.5 = (0.107)(28) + 2.5 \approx 5.5$$

Since $(H'/h)_{cr}$ is less than 7, we use the actual value, or $(H'/h)_{cr-S} = 5.5$. From Equation (4-54)

$$\left(\frac{H'}{h}\right)_{cr-R} = \left(\frac{H'}{h}\right)_{cr-S}[0.73 + 0.27(\frac{B}{h})] = 5.5[0.73 + 0.27(\frac{0.8}{0.4})] \approx 6.99$$

This value of 6.99 is less than $(1.55)(5.5) = 8.525$. So use $(H'/h)_{cr-R} = 6.99$. However, the actual $H'/h = 1.2/0.4 = 3$. So it is a shallow anchor. From Equation (4-57)

$$F'_{c-\psi=0°} = [7.46 + 1.44(\tfrac{h}{B})]\beta$$

Referring to Figure 4-21, for $(H'/h)/(H'/h)_{cr} = 3/6.99 = 0.429$, $\beta = 0.69$. So

$$F'_{c-\psi=0°} = [7.56 + (1.44)(\tfrac{0.4}{0.8})](0.69) = 5.71$$

Calculation of $F'_{c-\psi=90°}$

From Equation (4-59)

$$(\tfrac{H'}{h})_{cr-S} = 4.2 + 0.0606c_u = 4.2 + (0.0606)(28) = 5.9$$

So use $(H'/h)_{cr-S}$ as 5.9 since it is less than 6.5. From Equation (4-58)

$$\frac{(\tfrac{H'}{h})_{cr-R} + 0.5}{(\tfrac{H'}{h})_{cr-S} + 0.5} = [0.9 + 0.1(\tfrac{B}{h})] = 0.9 + (0.1)(2) = 1.1$$

$$(\tfrac{H'}{h})_{cr-R} + 0.5 = 1.1[(\tfrac{H'}{h})_{cr-S} + 0.5] = 1.1(5.9 + 0.5) = 7.04$$

So

$$(\tfrac{H'}{h})_{cr-R} = 6.54$$

For this case, since $(H'/h)_{cr-R} = 6.54 > H'/h = 1.2/0.4 = 3.0$, the anchor is shallow. Referring to Equation (4-62)

$$n' = \frac{(\tfrac{H'}{h}) + 0.5}{(\tfrac{H'}{h})_{cr} + 0.5} = \frac{3 + 0.5}{6.54 + 0.5} = 0.497$$

$$F^*_{c-\psi=90°} = 9[0.825 + 0.175(\tfrac{h}{B})] = 9[0.825 + 0.175(\tfrac{0.4}{0.8})] = 8.21$$

So, from Equation (4-61)

$$F'_{c-\psi=90°} = \frac{(8.21)(0.497)}{0.41 + (0.59)(0.497)} = 5.8$$

However [Equation (4-53)]

$$F'_{c-\psi} = F'_{c-\psi=0°} + (F'_{c-\psi=90°} \quad F'_{c-\psi=0°})(\frac{\psi°}{90})^2$$

$$= 5.71 + (5.8 \quad 5.71)(\frac{30}{89})^2 = 5.72$$

So, from Equation (4-51)

$$Q_u = Bh(F'_{c-\psi}c_u + \gamma H'\cos^2\phi)$$

$$= (0.8)(0.4)[(5.72)(28) + (18.4)(1.2)(\cos^2 30)] = \textbf{56.55 kN}$$

REFERENCES

Baker, W.H. and Kondner, R.L., 1966. Pullout load capacity of circular earth anchor buried in sand. *Highway Research Record No. 108*, National Academy of Sciences, Washington, D.C., 1-10.

Balla, A., 1961. The resistance of breakout of mushroom foundations for pylons. *Proc, V Intl. Conf. Soil Mech. Found. Engrg.*, Paris, 1:569-576.

Brinch Hansen, J., 1961. The ultimate resistance of rigid piles against transverse forces. *Bull. No. 12*, Danish Geotech. Inst., Copenhagen, Denmark.

Caquot, A. and Kerisel, L., 1949. *Traite de Mechanique des Sols*. Gauthier-Villars, Paris.

Das, B.M., 1985. Resistance of shallow inclined anchors in clay. In *Uplift Behavior of Anchor Found. in Soils* (Ed., S.P. Clemence), ASCE:86-101.

Das, B.M., 1980. A procedure for estimation of ultimate uplift capacity of foundations in clay. *Soils and Found.*, Japan, 20(1):72-82.

Das, B.M., Tarquin, A.J. and Moreno, R., 1985. Model tests for pullout resistance of vertical anchors in clay. *Civil Engrg. for Practicing and Design Engrs.*, Pergamon Press, 4(2):191-209.

Hanna, A.M., Das, B.M. and Foriero, A., 1988. Behavior of shallow inclined plate anchors in sand. In *Spec. Topics in Found.*, Geotech. Spec. Tech. Publ. No. 16, ASCE:54-72.

Harvey, R.C. and Burley, E., 1973. Behavior of shallow inclined anchorages in cohesionless sand. *Ground Engrg.*, 6(5):48-55.

Kanayan, A.S., 1966. Experimental investigation of the stability of bases of anchor foundations. *Soil Mech. Found. Engrg.*, Moscow, 3(6):9.

Maiah, A.A., Das, B.M. and Picornell, M., 1986. Ultimate resistance of shallow inclined strip anchor plate in sand. *Proc., Southeastern Conf. on Theoretical and Applied Mech.* (SECTAM XIII), Columbia, S.C., USA, 2:503-509.

Meyerhof, G.G. (1973). Uplift resistance of inclined anchors and piles. *Proc.*, VIII Intl. Conf. Soil Mech. Found. Engrg., Moscow, USSR, 2.1:167-172.

Meyerhof, G.G. and Adams, J.I, 1968. The ultimate uplift capacity of foundations. *Can. Geotech. J.*, 5(4):225-244.

Neely, W.J., Stuart, J.G. and Graham, J., 1973. Failure load of vertical anchor plates in sand. *J. Geotech. Engrg. Div.*, ASCE, 99(9):669-685.

Sokolovskii, V.V., 1965. *Statics of granular media.* Pergamon Press, London.

Chapter 5
HELICAL ANCHORS

5.1 INTRODUCTION

The design and use of helical anchors was briefly discussed in Chapter 1 (Section 1.4). Figures 1-8 and 1-9 show photographs of helical anchors with single and dual helices, which are generally used for light to medium loads. However, at the present time, tapered multi-helix anchors (3 to 4 helices) are commonly used to carry uplift loads up to about 550 kN. Figure 5-1 shows the typical dimensions of a multi-helix anchor used in the United States for construction of foundations of electrical transmission towers. These anchors are fairly easy to install and, hence, are cost effective.

At the present time only a few studies are available, the results of which can be used to estimate the ultimate uplift capacity of helical anchors. In many instances the ultimate load estimate is based on the rule of thumb. This chapter summarizes the existing theories relating to the prediction of the net ultimate uplift capacity of tapered multi-helix anchors embedded in sandy and clayey soils

HELICAL ANCHORS IN SAND

5.2 GEOMETRIC PARAMETERS AND FAILURE MODE

Figure 5-2 shows a tapered multi-helix anchor embedded in soil subjected to a vertical uplifting force. The diameter of the top helix is D_1 and that of the bottom helix is D_n. The distance between the ground surface and the top helix is H_1 and, similarly, the distance between the bottom helix and the ground surface is H_n The gross and net ultimate uplift capacities of the anchor can be expressed as

$$Q_{u(g)} = Q_u + W_a \qquad (5-1)$$

where $Q_{u(g)}$ = gross ultimate uplift capacity
Q_u = net ultimate uplift capacity
W_a = effective self-weight of the anchor

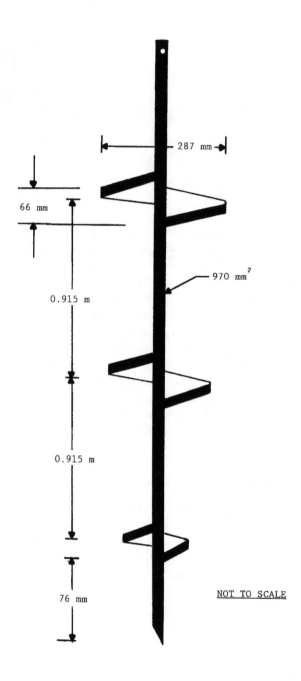

Figure 5-1   Typical multi-helix anchor used in the USA

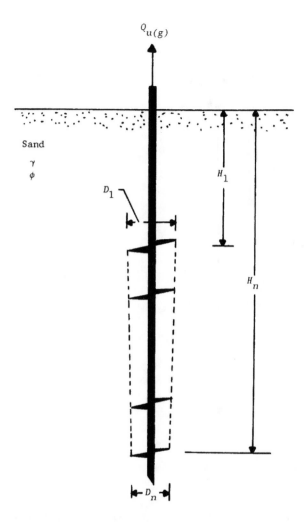

$Q_{u(g)}$

Sand

γ

ϕ

D_1

H_1

H_n

D_n

Figure 5-2 Tapered multi-helix anchor embedded in sand and subjected to uplift

Using laboratory model tests, Mitsch and Clemence (1985) studied the failure surface in soil around a helical anchor at ultimate load. Figure 5-3 shows a schematic diagram of the failure pattern for the condition where the embedment ratio H_1/D_1 is relatively small. For this case, it can be seen that:

1. The failure surface above the top helix is a truncated cone extending to the ground surface. The central angle of the truncated cone is approximately equal to the soil friction angle ϕ.

172

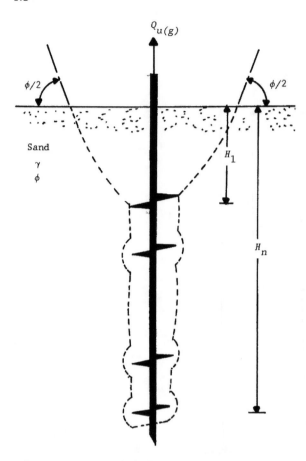

Figure 5-3   Typical failure pattern in sand around a multi-helix
anchor--shallow anchor condition

2.   Below the top helix, the failure surface is soil is approximately
cylindrical. This means that the inter-helical soil below the top
helix acts similar to a pile foundation with shear failure occurring
along the interface boundary.

When the conical failure surface of soil located above the top helix extends to the ground surface, it is referred to as shallow anchor condition.
However, if the anchor is located in such a way that H_1/D_1 is fairly large,
the failure surface in soil does not extend to the ground surface as shown in
Figure 5-4. This is referred to as deep anchor condition.

    In granular soils, the limiting value of $H_1/D_1 = (H_1/D_1)_{cr}$ at which the
anchor condition changes from shallow to deep is similar to that suggested by
Meyerhof and Adams (1968). Following are values of $(H_1/D_1)_{cr}$ for various soil
friction angles. These variations are also shown in Figure 5-5.

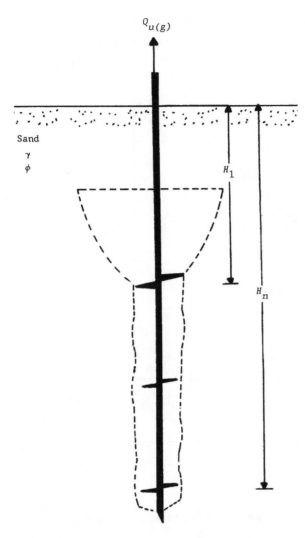

Figure 5-4   Typical failure pattern in sand around a multi-helix
anchor--deep anchor condition

Soil friction angle, ϕ (deg)	$(H_1/D_1)_{cr}$
25	3
30	4
35	5
40	7
45	9
48	11

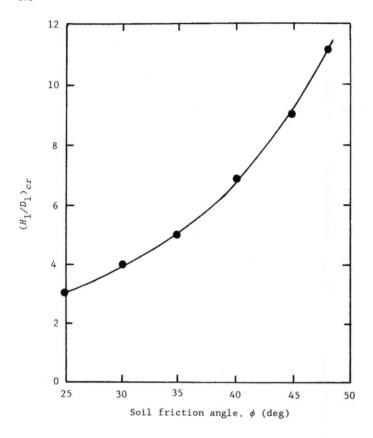

Figure 5-5   Variation of $(H_1/D_1)_{cr}$ with soil friction angle ϕ

5.3 NET ULTIMATE UPLIFT CAPACITY FOR SHALLOW ANCHOR CONDITION

Figure 5-6 shows an idealized failure surface in soil around a helical anchor at ultimate load. The net ultimate load can be approximately estimated according to the procedure outline by Mitsch and Clemence (1985), or

$$Q_u = Q_p + Q_f \tag{5-2}$$

where Q_p = bearing resistance for the top helix

Q_f = frictional resistance derived at the interface of the inter-helical soil which is cylindrical in shape

The magnitude of Q_p can be given as

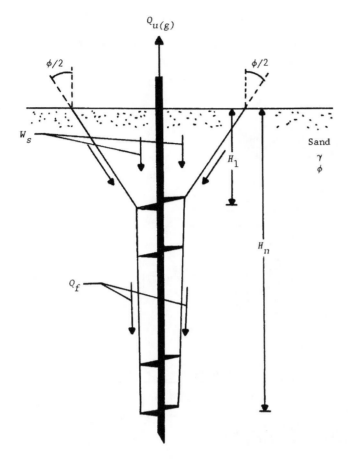

Figure 5-6 Idealized failure surface in sand for shallow anchor
condition--$H_1/D_1 \leq (H_1/D_1)_{cr}$

$$Q_p = \pi\gamma K_u(\tan\phi)[\cos^2(\tfrac{\phi}{2})]\left[\frac{D_1 H_1^2}{2} + \frac{H_1^3\tan(\tfrac{\phi}{2})}{3}\right] + W_s \qquad (5\text{-}3)$$

where γ = unit weight of soil

ϕ = soil friction angle

K_u = coefficient of lateral earth pressure in uplift

W_s = weight of the soil in the failure zone

The weight of the soil, W_s, can be expressed as

$$W_s = \gamma\left\{\frac{\pi}{3}H_1[(D_1^2) + (D_1 + 2H_1\tan\tfrac{\phi}{2})^2 + (D_1)(D_1 + 2H_1\tan\tfrac{\phi}{2})]\right\} \qquad (5\text{-}4)$$

The magnitude of Q_p can be expressed in a nondimensional form as

$$F_q = \frac{Q_p}{\gamma A H_1} \tag{5-5}$$

where $A = \frac{\pi}{4} D_1^2$

Now

$$\frac{\pi \gamma K_u (\tan \phi)[\cos^2(\frac{\phi}{2})]\left[\frac{D_1 H_1^2}{2} + \frac{H_1^3 \tan(\frac{\phi}{2})}{3}\right]}{\gamma A H_1}$$

$$= \frac{\pi \gamma K_u (\tan \phi)[\cos^2(\frac{\phi}{2})](H_1^3)\left[\frac{D_1}{2H_1} + \frac{\tan(\frac{\phi}{2})}{3}\right]}{\gamma(\frac{\pi}{4} D_1^2)H_1}$$

$$= 4K_u (\tan \phi)[\cos^2(\frac{\phi}{2})]\left(\frac{H_1}{D_1}\right)^2\left[\frac{0.5}{\left(\frac{H_1}{D_1}\right)} + 0.333 \tan(\frac{\phi}{2})\right] \tag{5-6}$$

Similarly

$$\frac{W_s}{\gamma A H_1} = \frac{\frac{\pi \gamma H_1}{3}[(D_1^2) + (D_1^2 + 4H_1^2 \tan^2 \frac{\phi}{2} + 4D_1 H_1 \tan \frac{\phi}{2}) + (D_1^2 + 2H_1 D_1 \tan \frac{\phi}{2})]}{\gamma(\frac{\pi}{4} D_1^2)H_1}$$

$$= 4 + 5.33\left(\frac{H_1}{D_1}\right)^2 \tan^2 \frac{\phi}{2} + 8\left(\frac{H_1}{D_1}\right)^2 \tan \frac{\phi}{2} \tag{5-7}$$

Let

$$\frac{H_1}{D_1} = G \tag{5-8}$$

Now, combining Equations (5-3), (5-5), (5-6), (5-7) and (5-8), we obtain

$$F_q = \frac{Q_p}{\gamma A H_1} = 4G^2 K_u (\tan \phi)[\cos^2(\frac{\phi}{2})][\frac{0.5}{G} + 0.333 \tan(\frac{\phi}{2})]$$

$$+ 4 + 5.33G^2 \tan^2(\frac{\phi}{2}) + 8G \tan(\frac{\phi}{2}) \tag{5.9}$$

In order to determine the breakout factor, F_q, we need to determine the magnitude of the coefficient of lateral earth pressure in uplift, K_u. The variation of K_u with soil friction angle ϕ suggested by Mitsch and Clemence (1985) can be expressed in the form

$$K_u = 0.6 + m\left(\frac{H_1}{D_1}\right)$$
(5-10)

where m = a coefficient which is a function of the soil friction angle ϕ. The variation of m is given in Table 5-1 and also in Figure 5-7

Table 5-1   Variation of m

Soil friction angle, ϕ (deg)	m
25	0.333
30	0.075
35	0.18
40	0.25
45	0.289

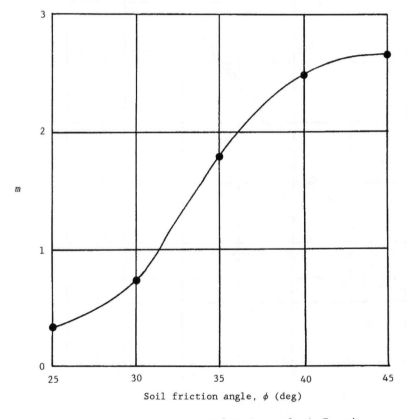

Figure 5-7   Variation of m with soil friction angle ϕ--Equation (5-10)

178

The magnitude of K_u increases with H_1/D_1 up to a maximum value and remains constant thereafter. This maximum value is attained at $(H_1/D_1)_{cr} = G_{cr}$ Based on this concept, the variations of K_u with H_1/D_1 and ϕ have been calculated and are shown in Figure 5-8. Substituting proper values of K_u and G into Equation (5-9), the variations of the breakout factor were calculated and are shown in Figure 5-9 and Table 5-2. Note that these plots are for $H_1/D_1 \leq (H_1/D_1)_{cr}$ So

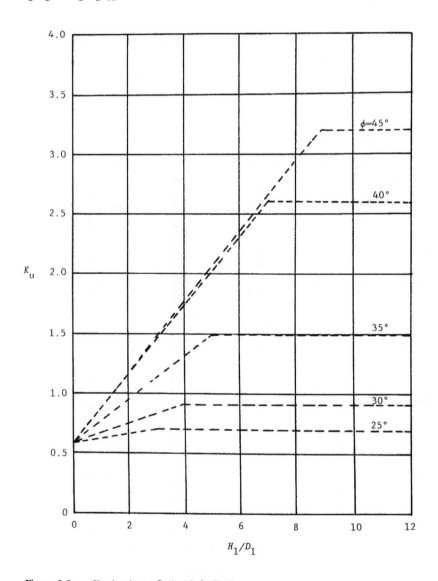

Figure 5-8   Variation of K_u with H_1/D_1

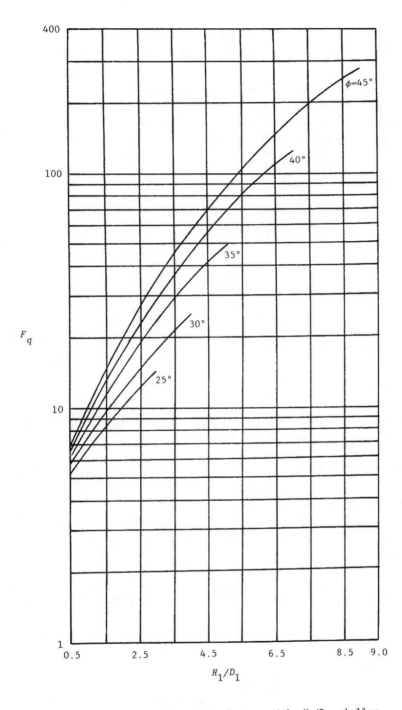

Figure 5-9   Variation of breakout factor with H_1/D_1--shallow condition

Table 5-2   Variation of Breakout Factor F_q for Shallow
Anchor Condition

	F_q				
H_1/D_1	$\phi = 25°$	$\phi = 30°$	$\phi = 35°$	$\phi = 40°$	$\phi = 45°$
0.5	5.27	5.54	5.87	6.23	6.61
1.0	6.74	7.38	8.25	9.18	10.17
1.5	8.41	9.54	11.16	12.91	14.77
2.0	10.27	12.01	14.64	17.49	20.53
2.5	12.33	14.82	18.72	22.99	27.54
3.0	14.60	17.97	23.44	29.46	25.91
3.5		21.48	28.84	36.99	45.74
4.0		25.35	34.95	45.64	57.13
4.5			41.81	55.44	70.18
5.0			49.46	66.56	85.00
5.5				78.97	101.68
6.0				92.76	120.34
6.5				108.01	141.06
7.0				124.78	163.98
7.5					189.14
8.0					216.69
8.5					246.73
9.0					279.34

$$Q_p = \frac{\pi}{4} F_q \gamma D_1^2 H_1 \tag{5-11}$$

Again, referring to Equation (5-2), the frictional resistance derived at the interface of the inter-helical soil can be given as

$$Q_f = \frac{\pi}{2} D_a \gamma (H_n^2 \quad H_1^2) K_u \tan \phi \tag{5-12}$$

where D_a = average helix diameter = $(D_1 + D_n)/2$

Thus, the net ultimate capacity can be given as [Equations (5-2), (5-3) and (5-12)]

$$Q_u = \pi \gamma K_u (\tan \phi)[\cos^2(\tfrac{\phi}{2})]\left[\frac{D_1 H_1^2}{2} + \frac{H_1^3 \tan(\tfrac{\phi}{2})}{3}\right] + W_s + (\tfrac{\pi}{2})(\frac{D_1 + D_n}{2})(\gamma) \cdot$$

$$(H_n^2 \quad H_1^2) K_u \tan \phi \tag{5-13}$$

or

$$Q_u = \underset{\uparrow}{\frac{\pi}{4} F_q \gamma D_1^2 H_1} \quad + (\tfrac{\pi}{2})(\frac{D_1 + D_n}{2})(\gamma)(H_n^2 \quad \underset{\uparrow}{H_1^2) K_u \tan \phi} \tag{5-14}$$

Equation (5-11)   Equation (5-12)

When using Equation (5-14), the following facts need to be kept in mind:

1. The breakout factors F_q shown in Figure 5-9 and Table 5-2 have been calculated using the values of K_u (as shown in Figure 5-8) for given H_1/D_1 ratios. Depending on the H_1/D_1 ratio of a given anchor, the magnitude of F_q can be picked from Figure 5-9 or Table 5-2 and used in Equation (5-14).

2. It is recommended that the K_u value to be used in Equation (5-12) [which is the second term in the right-hand side of Equation (5-14)] should be the maximum value (for the given friction angle). This means

$$K_u = 0.6 + m\left(\frac{H_1}{D_1}\right)_{cr}$$

Following are the maximum values of K_u for various soil friction angles ϕ.

Soil friction angle, ϕ (deg)	Maximum value of K_u
25	0.7
30	0.9
35	1.5
40	2.35
45	3.2

Example

5.1

Figure 5-10 shows a tapered multi-helix anchor. Given, for the sand: $\gamma = 102$ lb/ft^3, $\phi = 35°$ Given, for the anchor: $D_1 = 12$ in, $D_n = 7.5$ in, $H_1 = 3$ ft, $H_n = 10$ ft. Determine the net ultimate uplift capacity.

Solution

From Equation (5-14)

$$Q_u = \frac{\pi}{4}F_q \gamma D_1^2 H_1 + \left(\frac{\pi}{2}\right)\left(\left(\frac{D_1 + D_n}{2}\right)\right)(\gamma)(H_n^2 \quad H_1^2)K_u \tan \phi$$

From Table 5-2 for $\phi = 35°$, $H_1/D_1 = 3/1 = 3$, for which the magnitude of F_q is 23.44. Also for $\phi = 35°$, the maximum value of K_u is 1.5. So

182

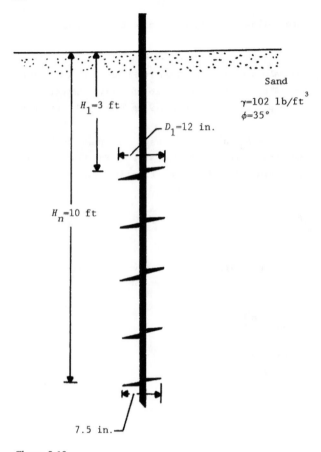

Figure 5-10

$$Q_u = (\frac{\pi}{4})(23.44)(102)(\frac{12}{12})^2(3) + (\frac{\pi}{2})[\frac{12 + 7.5}{(2)(12)}](102) \cdot$$

$$(10^2 \quad 3^2)(1.5)(\tan 35)$$

$$= 5633.4 + 12,442.4 \approx \mathbf{18,076 \ lb}$$

5.4 NET ULTIMATE UPLIFT CAPACITY FOR DEEP ANCHOR CONDITION

Figure 5-11 shows the idealized failure surface in soil around a deep helical anchor embedded in sand. For this condition (Mitsch and Clemence, 1985)

$$Q_u = Q_p + Q_f + Q_s \tag{5-15}$$

In the preceding equation Q_p and Q_f are, respectively, the bearing resistance of the top helix and the frictional resistance at the interface of the inter-

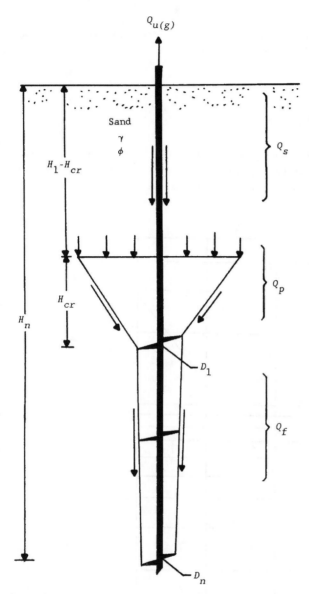

Figure 5-11 Idealized failure surface in sand for deep anchor condition

helical soil. The term Q_s is the frictional resistance derived from friction at the soil-anchor shaft interface above the top helix. It is recommended by the author that, due to various uncertainties involved in the determination of the soil parameters, the anchor shaft resistance Q_s may be neglected. So

$$Q_u \approx Q_p + Q_f \tag{5-16}$$

184

Bearing Resistance, Q_p

The bearing resistance Q_p of the top helix can easily be determined in terms of the breakout factor [as in Equation (5-11)], or

$$Q_p = \frac{\pi}{4}F_q^*\gamma D_1^2 H_1 \tag{5-17}$$

where F_q^* = deep anchor breakout factor

The magnitude of $F_q = F_q^*$ can be easily determined by substituting $G = G_{cr}$ and $K_u = K_{u(max)}$ in Equation (5-9). The variation of F_q^* has been calculated in this manner and plotted against the soil friction angle ϕ in Figure 5-12.

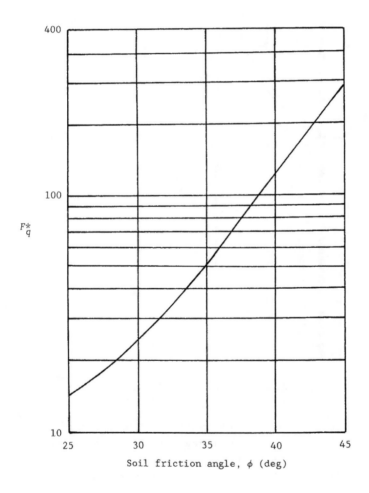

Figure 5-12   Variation of F_q^* with soil friction angle ϕ

Frictional Resistance, Q_f

The frictional resistance Q_f can be estimated by using Equation (5-12), or

$$Q_f = \frac{\pi}{2}D_a\gamma(H_n^2 \quad H_1^2)K_{u(max)}\tan\phi \qquad (5-18)$$

where

$$D_a = \frac{D_1 + D_n}{2} \qquad (5-19)$$

Net Ultimate Uplift Capacity

Equations (5-17), (5-18) and (5-19) can now be substituted into Equation (5-16) to obtain the net ultimate uplift capacity. So

$$Q_u = \frac{\pi}{4}F_q^*\gamma D_1^2 H_1 + (\frac{\pi}{2})(\frac{D_1 + D_n}{2})(\gamma)(H_n^2 \quad H_1^2)K_{u(max)}\tan\phi \qquad (5-20)$$

5.5 INTERFERENCE OF CLOSELY-SPACED ANCHORS

If helical anchors are placed too close to each other, the average net ultimate uplift capacity of each anchor may decrease due to the interference of the failure zones in soil located around the anchors. Laboratory model test results have shown that, for the non-interference of the anchor failure zones, the optimum center-to-center spacing in loose and dense sand should be $6D_1$ and $10D_1$, respectively. In any case, it is recommended that the minimum center-to-center spacing of the anchors should be about $5D_1$. A factor of safety of at least 2.5 should be used for estimation of the net allowable uplift capacity.

HELICAL ANCHORS IN CLAY ($\phi = 0$ CONDITION)

5.6 FAILURE MODE IN SOIL

Figure 5-13 shows a helical anchor embedded in a saturated clay having an undrained cohesion of c_u. If the H_1/D_1 ratio is relatively small (that is, *shallow anchor condition*) then, at ultimate load, the failure surface located above the top helix extends to the ground surface. Laboratory model test results of Mooney, Adamczak and Clemence (1985) showed that the nature of the maximum shear strain variation along the length of the anchor will be as shown. However if the H_1/D_1 ratio is relatively large the failure surface in soil above the top helix does not extend to the ground surface (that is, local shear failure takes place). This is referred to as *deep anchor condition*.

So, following the recommendations of Mooney et al. (1985), the idealized failure surfaces in soil for shallow and deep anchor conditions are shown in Figures 5-14a and 5-14b.

186

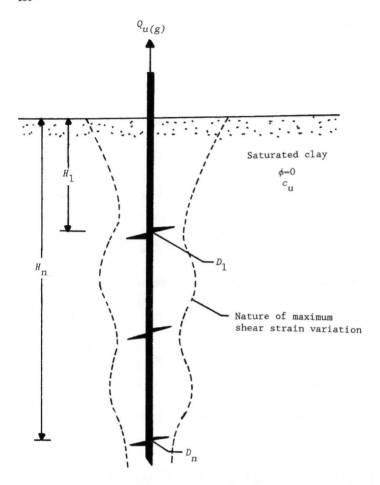

Figure 5-13   Failure mode in clay--shallow anchor condition

5.7   NET ULTIMATE UPLIFT CAPACITY

The net ultimate uplift capacity of the helical anchor can be given as [Equation (5-1)]

$$Q_u = Q_{u(g)} \quad W_a$$

For shallow anchors

$$Q_u = \underset{\underset{\text{Bearing resistance}}{\uparrow}}{Q_p} + \underset{\underset{\text{Resistance due to cohesion}}{\uparrow}}{Q_f} \qquad (5\text{-}21)$$

Bearing resistance   Resistance due to cohesion
of the top helix   at the interface of the
inter-helical soil

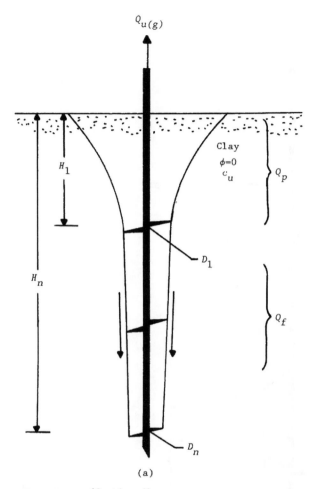

$Q_{u(g)}$

Clay
$\phi = 0$
c_u

Q_p

H_1

D_1

H_n

Q_f

D_n

(a)

Figure 5-14 (Continued)

Following the procedure of estimation of the uplift capacity of shallow plate anchors in clay, we can say that

$$Q_p = A(c_u F_c + \gamma H_1) \qquad (5\text{-}22)$$

where A = area of the top helix = $\frac{\pi}{4}(D_1^2)$

F_c = breakout factor

γ = unit weight of soil

H_1 = distance between the top helix and the ground surface

The magnitude of F_c increases with the H_1/D_1 ratio up to a maximum value of 9 at $(H_1/D_1)_{cr}$ The critical value of H_1/D_1 is a function of the undrained cohesion and can be expressed as (Das, 1980)

188

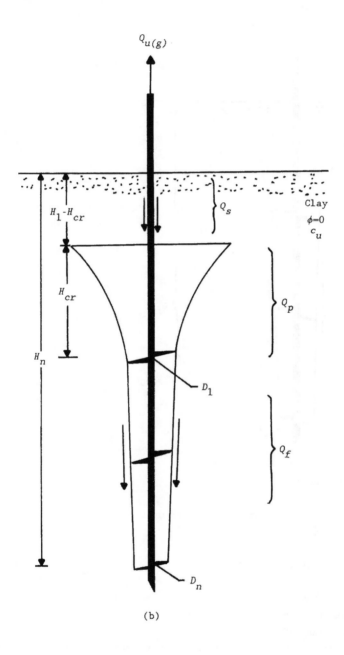

Figure 5-14 Idealized failure surface in clay at ultimate load:
(a) shallow condition; (b) deep condition

$$\left(\frac{H_1}{D_1}\right)_{cr} = 0.107c_u + 2.5 \leq 7 \tag{5-23}$$

where c_u is in kN/m^2

The variation of the breakout factor F_c can be estimated from Figure 5-15, which is a plot of F_c versus $(H_1/D_1)/(H_1/D_1)_{cr}$.

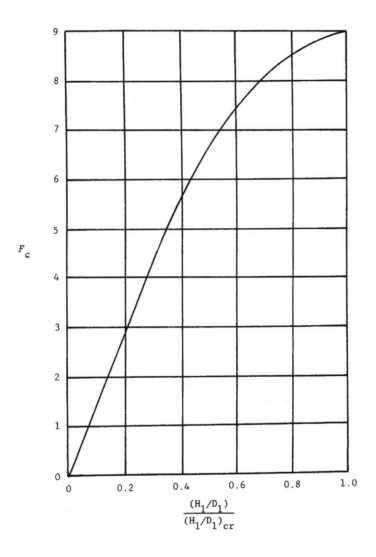

Figure 5-15 Variation of F_c with $(H_1/D_1)/(H_1/D_1)_{cr}$

The resistance due to cohesion at the interface of the inter-helical soil can be approximated as (Mooney et al. 1985)

$$Q_f = \pi(\frac{D_1 + D_n}{2})(H_n \quad H_1)c_u \tag{5-24}$$

Thus, combining Equations (5-21), (5-22) and (5-24), for *shallow anchor condition*

$$Q_u = \frac{\pi}{4}(D_1^2)(c_u F_c + \gamma H_1) + \pi(\frac{D_1 + D_n}{2})(H_n \quad H_1)c_u \tag{5-25}$$

In a similar manner, for *deep anchor condition*

$$Q_u = Q_p + Q_f + Q_s \tag{5-26}$$

where Q_s = resistance due to adhesion at the interface of the clay and the anchor shaft located above the top helix

The bearing resistance

$$Q_p = \frac{\pi}{4}(D_1^2)(9c_u + \gamma H_1) \tag{5-27}$$

The expression for Q_f will be the same as given for shallow anchor condition [Equation (5-24)]. The resistance due to adhesion at the interface of the clay and the anchor shaft located above the top helix can be approximated as

$$Q_s = p_s H_1 c_a \tag{5-28}$$

where p_s = perimeter of the anchor shaft
c_a = adhesion

The adhesion c_a may vary from about $0.3c_u$ for stiff clays to about $0.9c_u$ for very soft clays. Now, combining Equations (5-24), (5-26), (5-27) and (5-28), for deep anchor condition

$$Q_u = \frac{\pi}{4}(D_1^2)(9c_u + \gamma H_1) + \pi(\frac{D_1 + D_n}{2})(H_n \quad H_1)c_u + p_s H_1 c_a \tag{5-29}$$

In all cases, a factor of safety of at least 2.5 is recommended for determination of the net allowable uplift capacity.

Example

5.2
_____

Consider a multi-helix anchor embedded in a saturated clay. Given:

For the clay: $\gamma = 18.5$ kN/m^3

$c_u = 35$ kN/m^2

For the anchor: $D_1 = 0.4$ m; $D_n = 0.25$ m

$H_1 = 3$ m; $H_n = 7$ m

Diameter of the anchor shaft = 50 mm

Estimate the net ultimate uplift capacity.

Solution

$H_1 = 3$ m; $D_1 = 0.4$ m

$$\frac{H_1}{D_1} = \frac{3}{0.4} = 7.5$$

From Equation (5-23) it can be seen that the maximum value of $(H_1/D_1)_{cr}$ is 7 Since the H_1/D_1 is 7.5, it is deep anchor condition. So, from Equation (5-29)

$$Q_u = \frac{\pi}{4}(D_1^2)(9c_u + \gamma H_1) + \pi(\frac{D_1 + D_n}{2})(H_n \quad H_1)c_u + p_s H_1 c_a$$

$$p_s = (\pi)(\frac{50}{1000}) = 0.157 \text{ m}$$

Assume $c_a \approx 0.5c_u = (0.5)(35) = 17.5$ kN/m^2. So

$$Q_u = (\frac{\pi}{4})(0.4)^2[(9)(35) + (18.5)(3)] + (\pi)[\frac{0.4 + 0.25}{2}](7 \quad 3)(35)$$

$$+ (0.157)(3)(17.5)$$

$$= 47.6 + 142.9 + 8.24 = \textbf{197.74 kN}$$

_____

5.8 USE OF *IN-SITU* TESTS TO PREDICT UPLIFT PERFORMANCE

Lutenegger, Smith and Kabir (1988) conducted various types of in situ tests to determine the soil shear strength parameters for prediction of the net ultimate uplift capacity of multi-helix anchors. These tests included the following:

a. electric cone penetrometer

b. piezocone penetrometer

c. Marchetti dilatometer

d. borehole shear test

192

e. Pencel pressuremeter

f. Menard 3-cell and mono-cell pressuremeters

g. vane shear

Based on their tests, Lutenegger et al. concluded that, in sand and clay, the best results were obtained from the cone penetrometer and dilatometer tests.

REFERENCES

Mitsch, M.P. and Clemence, S.P., 1985. The uplift capacity of helix anchors in sand. *Proc.*, Uplift Behavior of Anchor Foundations in Soil (Ed. S.P. Clemence), ASCE:26-47.

Mooney, J.S., Adamczak, S., Jr. and Clemence, S.P., 1985. Uplift capacity of helix anchors in clay and silt. *Proc.*, Uplift Behavior of Anchor Foundations in Soil (Ed. S.P. Clemence), ASCE:48-72.

Meyerhof, G.G. and Adams, J.I., 1968. The ultimate uplift capacity of foundations. *Can. Geotech.. J.*, 5(4):225-244.

Das, B.M., 1980. A procedure for estimation of ultimate uplift capacity of foundations in clay. *Soils and Found.*, Japan, 20(1):77-82.

Lutenegger, A.J., Smith, B.L. and Kabir, M.B., 1988. Use of *in situ* tests to predict uplift performance of multi-helix anchors. In *Spec. Topics in Found.*, Geotech. Spec. Tech. Pub. 16 (Ed. B.M. Das), ASCE:93-110.

Chapter 6
ANCHOR PILES

6.1   INTRODUCTION

In the construction of various types of foundations, piles are generally used to transmit downwardly-directed load to a stronger soil at a greater depth. They are also used to resist lateral load imposed on a foundation. During the last three to four decades, several theoretical and experimental studies were conducted by various investigators to evaluate the downwardly-directed and lateral load-bearing capacity of single and group piles embedded in sandy and clayey soils.

Piles can also be used in the construction of foundations subjected to uplifting forces. The uplift force is resisted by *skin friction* developed at the soil-pile interface (Figure 6-1). At the present time, limited studies are available to estimate the uplift capacity of piles. Only a few laboratory model study results are available regarding the efficiency of group piles subjected to uplifting forces.

The net ultimate uplift capacity of a single pile can be expressed as (Figure 6-1)

$$Q_{u(g)} = Q_u + W_p \qquad\qquad (6\text{-}1)$$

where $Q_{u(g)}$ = gross ultimate uplift capacity
   Q_u = net ultimate uplift capacity
   W_p = effective self-weight of the pile

The net ultimate uplift capacity of a pile embedded in sand is primarily a function of the following parameters:

a.   length of embedment, L;

b.   pile diameter, D;

c.   roughness of the pile surface;

d.   soil friction angle ϕ and its relative density; and

e.   nature of placement of the pile (driven, bored or cast-in-place).

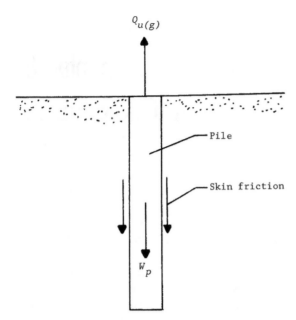

Figure 6-1   Pile subjected to uplifting load

In a similar manner, the magnitude of Q_u for a pile embedded in saturated or near-saturated clay is a function of

    a.   the length of embedment, L;

    b.   pile diameter, D;

    c.   undrained cohesion of the clay, c_u; and

    d.   the nature placement of pile.

In any case, it has been shown from laboratory model tests that the be-havior of a pile to uplift is somewhat different when compared to that of a plate anchor. This can be explained by referring to Figure 6-2, in which a pile of diameter D and a plate anchor of diameter h are embedded in a similar soil at the same depth below the ground surface (that is, L). For this case D = h. If both the pile and the plate anchor are subjected to uplifting force, the nature of the net load (Q) and the uplift movement (Δ) diagrams will be like those in Figure 6-3. A comparison of these two Q versus Δ plots shows that, under similar conditions, (a) the net ultimate uplift load of the pile is somewhat lower than that of the horizontal plate anchor; and (b) at ul-timate load, the ratio of Δ/D is relatively less for the pile compared to the Δ/h ratio of the plate anchor.

As is the case of other chapters, this chapter has been divided into two major parts: (a) piles in sand and (b) piles in saturated or near-saturated clay.

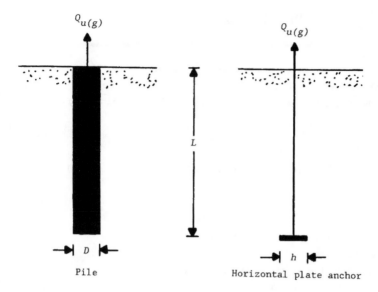

Figure 6-2

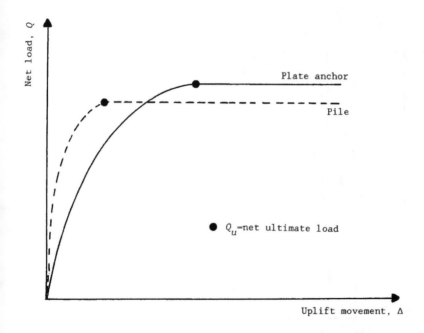

Figure 6-3 Comparison of the nature of net load versus uplift movement diagrams for pile and horizontal plate anchor

PILES IN SAND

6.2 BORED PILES

Figure 6-4 shows a vertical pile embedded in a granular soil. The length of embedment and the diameter of the pile are L and D, respectively. From Equation (6-1), the net ultimate uplift capacity of the pile can be given as

$$Q_u = Q_{u(g)} \quad W_p \qquad (6-2)$$

The net ultimate uplift capacity can be expressed as (Meyerhof, 1973a)

$$Q_u = (\sigma_o' K_u \tan \delta) A_s \qquad (6-3)$$

where σ_o' = average effective overburden pressure

K_u = uplift coefficient

δ = angle of friction at the soil-pile interface

A_s = embedded pile surface area

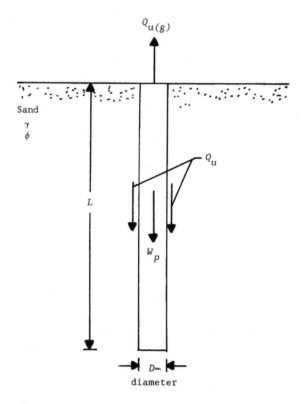

Figure 6-4 Bored pile in sand

The average effective vertical stress is

$$\sigma'_o = \frac{1}{2}\gamma L \qquad (6\text{-}4)$$

where γ = unit weight of sand

In the case of submerged sand, the unit weight γ in Equation (6-4) should be replaced by γ' (where $\gamma' = \gamma_{sat} \quad \gamma_w$; γ_w = unit weight of water). For piles having circular cross section, the embedded surface area is

$$A_s = \pi D L \qquad (6\text{-}5)$$

Thus, combining Equations (6-3), (6-4) and (6-5)

$$Q_u = (\frac{1}{2}\gamma L)(K_u)(\tan \delta)(\pi D L) = \frac{\pi}{2}\gamma D L^2 K_u \tan \delta \qquad (6\text{-}6)$$

The preceding relationship can also be expressed as

$$Q_u = A_s \overline{f} \qquad (6\text{-}7)$$

where $\overline{f}$ = average friction resistance per unit area of the soil-pile interface
Thus

$$\overline{f} = \frac{1}{2}K_u \gamma L \tan \delta \qquad (6\text{-}8)$$

The variation of the uplift coefficient K_u with the soil friction angle ϕ suggested by Meyerhof (1973a) is shown in Figure 6-5.

Das, Seeley and Pfeifle (1977) and Das (1983) provided results of several laboratory model tests for the ultimate uplift capacity of *rough piles* embedded in sand. These model tests were conducted to obtain the variation of the *frictional resistance per unit area*, f, along the embedded length of the pile. Based on Equation (6-8) it is obvious that

$$f = \gamma z K_u \tan \delta \qquad (6\text{-}9)$$

where z = distance measured from the ground surface (Figure 6-4)

The model tests of Das et al. (1977) and Das (1983) were conducted by changing the length of embedment L in small increments and then determining the net ultimate uplift capacity for each case. The frictional resistance per unit area f at a depth $z = (L_1 + L_2)/2$ (as shown in Figure 6-6) was calculated as

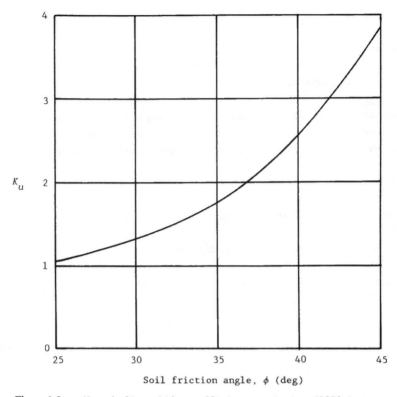

Figure 6-5   Meyerhof's uplift coefficient variation (1973a)

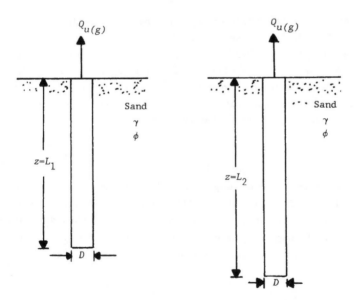

Figure 6-6   Determination of frictional resistance per unit area
(*f*)--Equation (6-10)

$$f = \frac{Q_{u(z=L_2)} \quad Q_{u(z=L_1)}}{p(L_1 \quad L_2)} \tag{6-10}$$

where $Q_{u(z=L_2)}$ = net ultimate uplift capacity of pile having an embedment

length $L_2 = Q_{u(g)(z=L_2)} \quad W_{p(L_2)}$

$Q_{u(z=L_1)}$ = net ultimate uplift capacity of pile having an embedment

length $L_1 = Q_{u(g)(z=L_1)} \quad W_{p(L_1)}$

p = perimeter of the cross section of the pile

The results of these tests show that, for a given soil and relative density of compaction, the frictional resistance f increases linearly with L/D up to a certain value and remains constant thereafter (Figure 6-7). The embedment ratio at which the magnitude of f becomes constant can be referred to as the critical embedment ratio, $(L/D)_{cr}$. The magnitude of the critical embedment ratio can be given by the following empirical relationships (Das, 1983)

$$(\tfrac{L}{D})_{cr} = 0.156D_r + 3.58 \quad (\text{for } D_r \leq 70\%) \tag{6-11}$$

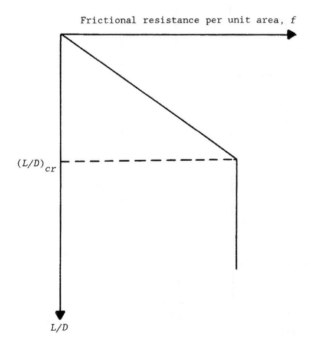

Figure 6-7   Nature of variation of L/D versus f

200

$$\left(\frac{L}{D}\right)_{cr} = 14.5 \quad \text{(for } D_r \geq 70\%) \tag{6-12}$$

where D_r = relative density of sand (in per cent)

Figure 6-8 shows a plot of $(L/D)_{cr}$ against D_r based on Equations (6-11) and (6-12).

The nature of variation of f with L/D shown in Figure 6-7 is similar to that observed by Vesic (1970) for piles under compressive load. Based on the model test results, Das et al. (1977) also provided the variation of δ/ϕ with D_r for rough bored piles, as shown in Figure 6-9.

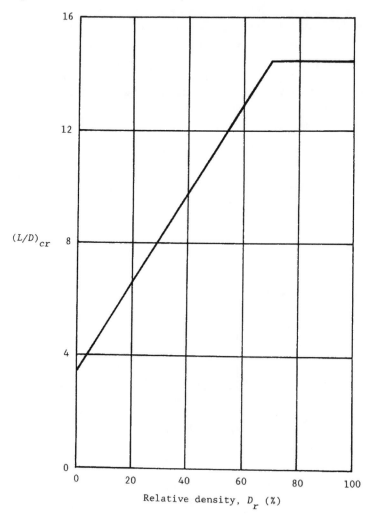

Figure 6-8   Plot of $(L/D)_{cr}$ versus D_r [Equations (6-11) and (6-12)]

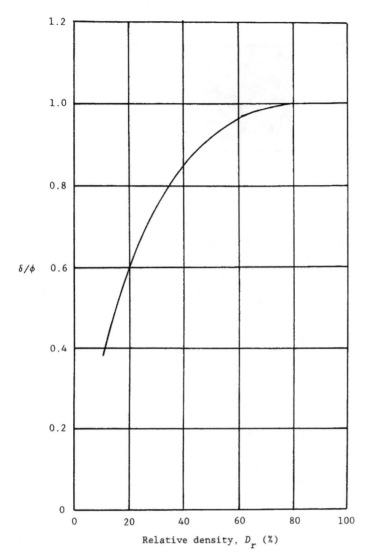

Figure 6-9 Variation of δ/ϕ with D_r based on the model test results of Das et al. (1977)

With the above theoretical and experimental results, it is now possible to develop the following step-by-step procedure for estimation of the net ultimate uplift capacity of *rough bored piles*.

1. For a given pile, obtain the embedment ratio L/D. (Note: $D =$ diameter of a pile having a circular cross section, and $D =$ length of each side of a pile having a square cross section as shown in Figure 6-10).

Circular pile   Square pile

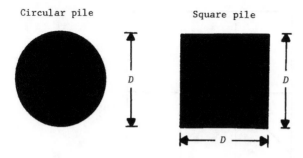

Figure 6-10

2. Estimate the relative density of compaction (D_r) of sand.

3. Determine the critical embedment ratio from Equations (6-11) and (6-12) (or Figure 6-8).

4. Compare the L/D ratio obtained in Step 1 with the $(L/D)_{cr}$ calculated in Step 3. If $L/D \leq (L/D)_{cr}$, go to Step 5. However, if $L/D > (L/D)_{cr}$, then go to Step 6.

5. If $L/D \leq (L/D)_{cr}$, then

$$Q_u = \int_0^L p \cdot f \cdot dz = \frac{1}{2} p \gamma L^2 K_u \tan \delta$$

For piles with circular cross section

$$Q_u = \frac{\pi}{2} D \gamma L^2 K_u \tan \delta \tag{6-13}$$

and for piles with square cross section

$$Q_u = (\frac{1}{2})(4D) \gamma L^2 K_u \tan \delta = 2D \gamma L^2 K_u \tan \delta \tag{6-14}$$

Knowing the value of ϕ, the magnitudes of K_u and δ can be determined from Figures 6-5 and 6-9, respectively.

6. If $L/D > (L/D)_{cr}$, then determine L_{cr}

$$L_{cr} = (\frac{L}{D})_{cr} \cdot D \tag{6-15}$$

$$Q_u = \int_0^{L_{cr}} p \cdot f \cdot dz + p \cdot f_{(at\ z=L_{cr})} \cdot (L \quad L_{cr}) \tag{6-16}$$

However (Figure 6-11)

$$f = \gamma z K_u \tan \delta \tag{6.17}$$

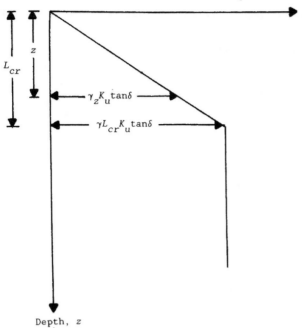

Frictional resistance per unit area, f

Depth, z

Figure 6-11

$$f_{(at\ z=L_{cr})} = \gamma L_{cr} K_u \tan \delta \qquad (6\text{-}18)$$

Substitution of Equations (6-17) and (6-18) into Equation (6-16) yields

$$Q_u = \frac{1}{2} p \gamma L_{cr}^2 K_u \tan \delta + p \gamma K_u \tan \delta (L \quad L_{cr}) \qquad (6\text{-}19)$$

So, for piles with circular cross section

$$Q_u = \frac{\pi}{2} D \gamma L_{cr}^2 K_u \tan \delta + \pi D \gamma K_u \tan \delta (L \quad L_{cr}) \qquad (6\text{-}20)$$

Similarly, for piles with square cross section

$$Q_u = 2 D \gamma L_{cr}^2 K_u \tan \delta + 4 D \gamma K_u \tan \delta (L \quad L_{cr}) \qquad (6\text{-}21)$$

Example

6.1

Consider a pile having a circular cross section with a diameter $D = 0.4$ m and length of embedment $L = 10$ m (Figure 6-12). Given, for the

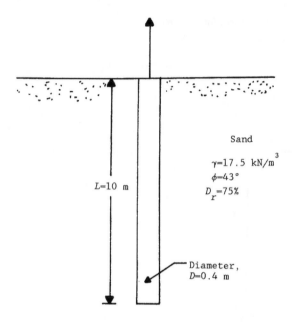

Figure 6-12

sand: $\phi = 43°$, $D_r = 75\%$, $\gamma = 17.5$ kN/m^3. Determine the net ultimate uplift capacity.

Solution

$D = 0.4$ m; $L = 10$ m; $L/D = 10/0.4 = 25$; $D_r = 75\%$ (given). Since $D_r = 75\%$, from Equation (6-12), $(L/D)_{cr} = 14.5$. For this pile $L/D > (L/D)_{cr}$, so Equation (6-20) will apply.

$$Q_u = \frac{\pi}{2}D\gamma L_{cr}^2 K_u \tan \delta + \pi D \gamma K_u \tan \delta (L \quad L_{cr})$$

$$L_{cr} = (\tfrac{L}{D})_{cr} \cdot D = (14.5)(0.4) = 5.8 \text{ m}$$

From Figure 6-5, for $\phi = 43°$, $K_u \approx 3.35$. Again from Figure 6-9, for $D_r = 75\%$, $\delta/\phi = 1$. So $\delta = 43°$ Now

$$Q_u = (\tfrac{\pi}{2})(0.4)(17.5)(5.8)(3.35)(\tan 43)$$

$$+ (\pi)(0.4)(17.5)(3.35)(\tan 43)(10 \quad 5.8) = 199.2 + 288.5$$

$$= \mathbf{487.7 \text{ kN}}$$

6.3 DRIVEN PILES

If piles are *driven* into the ground, the lateral earth pressure coefficient
will change when compared to the case of *bored piles*. This change of lateral
earth pressure coefficient at the pile-sand interface will cause a change in
the skin friction f at any given depth z measured from the ground surface.
For *rigid rough circular piles* embedded in sand, Meyerhof (1973b) proposed
that

$$Q_u = \frac{1}{2}\gamma L^2 DK'_u \qquad\qquad (6\text{-}22a)$$

where L = length of embedment

D = *diameter* of the circular pile

K'_u = modified uplift coefficient

The variations of K'_u with the soil friction angle ϕ proposed by Meyerhof
(1973b) are given in Figure 6-13. It is important to keep in mind the following facts while using Equation (6.22a) and Figure 6-13.

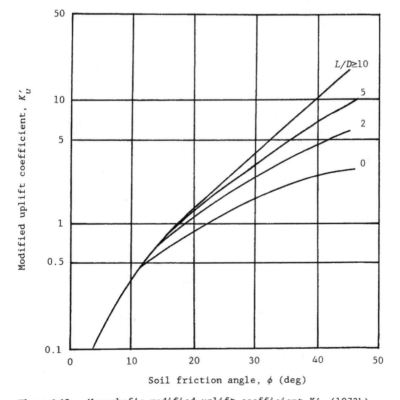

Figure 6-13   Meyerhof's modified uplift coefficient K'_u (1973b)

1. Equation (6-22a) is for *circular* piles only. For piles with square cross section

$$Q_u = \frac{1}{2}\gamma L^2 D(\frac{4}{\pi})K_u' = 0.637\gamma L^2 DK_u' \qquad (6\text{-}22b)$$

2. Presently no model or field test results for driven piles in sand are available to predict the magnitude of the critical embedment ratio $(L/D)_{cr}$ at which the unit skin friction f becomes constant as shown in Figure 6-7. For that reason, Equations (6-22a) and (6-22b) (and Figure 6-13) should be used for $L/D \leq 15$.

Example

6.2

Consider the pile described in Example Problem 6.1. The only change is that L is now equal to 6 m. Estimate the net ultimate uplift capacity by assuming that

a. it is a bored pile, and
b. it is a driven pile.

Solution

Given: $L = 6$ m, $D = 0.4$ m; $L/D = 15$; $\phi = 43°$; $\gamma = 17.5$ kN/m^3

Part a. From Equation (6-20)

$$Q_u = \frac{\pi}{2}D\gamma L_{cr}^2 K_u \tan \delta + \pi D\gamma K_u \tan \delta(L \quad L_{cr})$$

As in Example Problem 6.1, K_u 3.35; $L_{cr} = 5.8$. So

$$Q_u = (\frac{\pi}{2})(0.4)(17.5)(5.8)^2(3.35)(\tan 43)$$
$$+ (\pi)(0.4)(17.5)93.35)(\tan 43)(6 \quad 5.8) = 1155.5 + 13.7$$
$$= \mathbf{1169.2\ kN}$$

Part b. From Equation (6-22a)

$$Q_u = \frac{1}{2}\gamma L^2 DK_u'$$

From Figure 6-13, for $\phi = 43°$ the magnitude of K_u' is about 13.7. So

$$Q_u = (\frac{1}{2})(17.5)(6)^2(0.4)(13.7) \approx \mathbf{1726\ kN}$$

Comments: Comparing the results of Parts a and b, it can be seen that Q_u is about 50% higher for a driven pile compared to a bored pile.

6.4 UPLIFT CAPACITY OF INCLINED PILES SUBJECTED TO AXIAL PULL

Figure 6-14 shows a *rough rigid inclined pile* embedded in a sand having a unit weight γ and a friction angle ϕ. The length of the embedded inclined pile is equal to L. The inclination of the pile with respect to the vertical is α. The gross ultimate uplift capacity of an inclined pile can be given as

$$Q_{u(g)\alpha} = Q_{u\alpha} + W_p \cos \alpha$$

where $Q_{u(g)\alpha}$ = gross ultimate axial uplift capacity
$\quad Q_{u\alpha}$ = net ultimate axial uplift capacity
$\quad W_p$ = effective self-weight of the pile

Thus

$$Q_{u\alpha} = Q_{u(g)\alpha} \quad W_p \cos \alpha \qquad (6\text{-}23)$$

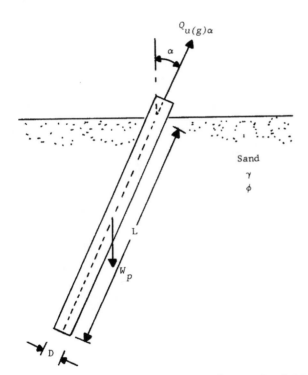

Figure 6-14   Geometric parameters of a rough rigid inclined pile embedded in sand

Based on laboratory model test results, Tran-Vo-Nhiem (1971) proposed that

$$Q_{u\alpha} = \frac{Q'_{u(\alpha=0)}}{\cos \alpha} \quad \text{(for } \alpha \leq 40°) \tag{6-24}$$

where $Q'_{u(\alpha=0)}$ = net ultimate pullout resistance of a similar pile having a length $L' = L \cos \alpha$

Meyerhof (1973a) also proposed a theoretical relationship for $Q_{u(\alpha)}$ for *bored piles* as

$$Q_{u\alpha} = [\sigma'_o K_{u\alpha} \tan \delta] A_s \tag{6-25}$$

where σ'_o = average effective overburden pressure

$K_{u\alpha}$ = uplift coefficient of inclined piles

δ = angle of friction at the soil-pile interface

A_s = embedded pile surface area

For this problem, referring to Figure 6-15

$$\sigma'_o = (\frac{L'}{2})\gamma = \frac{1}{2} \gamma L \cos \alpha \tag{6-26}$$

For *circular piles*

$$A_s = pL = \pi DL \tag{6-27}$$

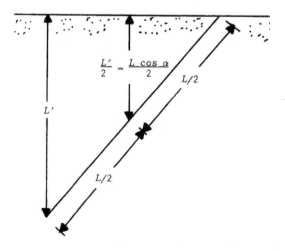

Figure 6-15

Thus, combining Equations (6-25), (6-26) and (6-27)

$$Q_{u\alpha} = \frac{\pi}{2}\gamma L^2 D K_{u\alpha} \cos \alpha \cdot \tan \delta \qquad (6\text{-}28)$$

However, for piles with *square* cross section

$$A_s = pL = 4DL \qquad (6\text{-}29)$$

So, from Equations (6-25), (6-26) and (6-29)

$$Q_{u\alpha} = 2\gamma L^2 D K_{u\alpha} \cos \alpha \cdot \tan \delta \qquad (6\text{-}30)$$

The variation of the uplift coefficient for inclined piles $[K_{u(\alpha)}]$ proposed by Meyerhof (1973a) is shown in Figure 6-16. Although no theoretical or experimental results are available at this time, the author feels that some modifications in Equations (6-28) and (6-30) are necessary in view of the fact

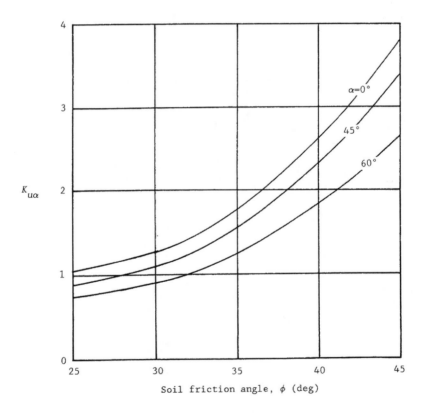

Figure 6-16   Meyerhof's values of $K_{u\alpha}$ (1973a)

210

that the unit friction f at the soil-pile interface may become constant beyond a certain depth as shown in Figure 6-17 (similar to that for vertical piles as shown in Figure 6-7; Section 6.2). This modification will result in a conservative value of $Q_{u\alpha}$. The proposed modification can be done by substituting L' in Equations (6-11) and (6-12) for the critical embedment ratio, or

$$(\frac{L'}{D})_{cr} = (\frac{L\cos\alpha}{D})_{cr} = 0.156D_r + 3.58 \quad \text{(for } D_r \leq 70\%) \quad (6\text{-}31a)$$

and

$$(\frac{L'}{D})_{cr} = (\frac{L\cos\alpha}{D})_{cr} = 14.5 \quad \text{(for } D_r > 70\%) \quad (6\text{-}31b)$$

Proceeding a manner similar to that shown in Section 6.2, we can thus obtain the following relationships.

Case 1. For $L' \leq (\frac{L'}{D})_{cr} \cdot D = L'_{cr}$

$$Q_{u\alpha} = \frac{\pi}{2}D\gamma L'^2 K_{u\alpha}\cos\alpha \cdot \tan\delta \quad \text{(for circular piles)} \quad (6\text{-}32)$$

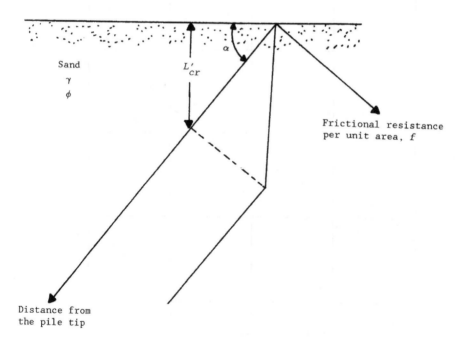

Figure 6-17   Variation of frictional resistance per unit area f along the surface of the inclined pile

and

$$Q_{u\alpha} = 2D\gamma L^2 K_{u\alpha} \cos \alpha \cdot \tan \delta \quad \text{(for square piles)} \tag{6-33}$$

Case 2. For $L' > (\frac{L'}{D})_{cr} \cdot D = L'_{cr}$

$$Q_{u\alpha} = \frac{\pi}{2} D\gamma L^2_{cr} K_{u\alpha} \cos \alpha \cdot \tan \delta + \pi D\gamma K_{u\alpha} \cos \alpha \cdot \tan \alpha$$

$$\text{(for circular piles)} \tag{6-34}$$

and

$$Q_{u\alpha} = 2D\gamma L^2_{cr} K_{u\alpha} \cos \alpha \cdot \tan \delta + \pi D\gamma K_{u\alpha} \cos \alpha \cdot \tan \delta (L \quad L_{cr})$$

$$\text{(for square piles)} \tag{6-35}$$

In Equations (6-34) and (6-35)

$$L_{cr} = \frac{L'_{cr}}{\cos\alpha} \tag{6-36}$$

In Equations (6-32), (6-33), (6-34) and (6-35), the magnitudes of $K_{u(\alpha)}$ and δ should be obtained from Figures 6-16 and 6-9, respectively.

General Comments

In light of the above derivations, it is essential to keep in mind the following general facts while estimating the net ultimate uplift capacity of inclined piles subjected to axial pullout force.

1. Equation (6-24) is recommended for use with the pile inclination α being equal to or less than 40 degrees and L/D being equal to or less than about 20. If this empirical relationship is used for bored piles, then $Q'_{u(\alpha=0)}$ needs to be determined by using the proper equation [Equation (6-13), (6-14), (6-20) or (6-21)]. However, for driven piles, $Q_{u(\alpha=0)}$ needs to be determined using Equation (6-22a) or (6-22b). In addition note that, for all calculations in Equations (6-13), (6-14), (6-20), (6-21), (6-22a) and (6-22b) (if applicable), $L' = L \cos \alpha$ should be used in place of L.

2. Equations (6-32), (6-33), (6-34) and (6-35) are only applicable for *bored piles*.

212

Example

6.3

Consider the inclined bored pile embedded in sand shown in Figure 6-18. Determine the net axial uplift capacity $Q_{u\alpha}$ using Equation (6-24).

Solution

For this problem, given: $L = 10$ ft, $D = 1$ ft, $\gamma = 108$ lb/ft^3, $\phi = 35°$, $D_r = 60\%$ and $\alpha = 40°$ According to Equation (6-24)

$$Q_{u\alpha} \approx \frac{Q'_{u(\alpha=0)}}{\cos 40}$$

In order to estimate $Q'_{u(\alpha=0)}$ we need to determine which of the two equations should be used [either Equation (6-13) or (6-20)]. For this case

$$\frac{L'}{D} = \frac{L \cos \alpha}{D} = \frac{(10)(\cos 40)}{1} = 7.66$$

From Equation (6-11)

$$\left(\frac{L'}{D}\right)_{cr} = 0.156D_r + 3.58 = (0.156)(60) + 3.58 = 12.94$$

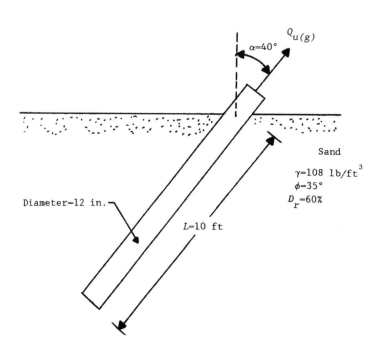

Figure 6-18

Since $L'/D < (L'/D)_{cr}$, Equation (6-13) will be used with L being replaced by L'. Thus

$$Q_{u(\alpha=0)} = \frac{\pi}{2} D \gamma L'^2 K_u \tan \delta$$

From Figure 6-5 for $\phi = 35°$, the value of K_u is about 1.83. Also from Figure 6-9, for $D_r = 60\%$, the magnitude of δ/ϕ is about 0.97. So $\delta = (0.97)(35) = 33.95°$. So

$$Q_{u(\alpha=0)} = (\frac{\pi}{2})(1)(108)(10 \cdot \cos40)^2 (1.83)(\tan33.95) \approx 12,265 \text{ lb}$$

$$\approx 12.3 \text{ kips}$$

So

$$Q_{u\alpha} = \frac{12.3}{\cos 40} = \textbf{16.06 kips}$$

**Example
6.4**

Solve Example Problem 6.3 assuming that the pile is driven.

Solution

From Equation (6-22a)

$$Q'_{u(\alpha=0)} = \frac{1}{2} \gamma L'^2 D K'_u$$

$L' = (10)(\cos 40) = 7.66$. For $L'/D = 7.66$ and $\phi = 35°$, the value of K'_u (Figure 6-13) is about 5. So

$$Q'_u = (\frac{1}{2})(108)(7.66)^2 (1)(5) = 15,842 \text{ lb} \approx 15.8 \text{ kips}$$

So

$$Q_{u\alpha} = \frac{15.8}{\cos 40} \approx \textbf{20.63 kips}$$

**Example
6.5**

Solve Example Problem 6.3 assuming the pile is a bored pile and using either Equation (6-32) or (6-34), whichever is applicable.

Solution

From Equation (6-31a)

$$(\frac{L'}{D})_{cr} = 0.156 D_r + 3.58 = (0.156)(60) + 3.58 = 12.94$$

For the present problem

$$L' = (L)(\cos 40) = (10)(\cos 40) = 7.66$$

Since $L' < L_{cr}$, Equation (6-32) will apply. Thus

$$Q_{u\alpha} = \frac{\pi}{2} D \gamma L^2 K_{u\alpha} \cos \alpha \cdot \tan \delta$$

As in Example Problem 6.3, $\delta = 33.95°$. From Figure 6-16, for $\phi = 35°$ and $\alpha = 40°$, the value of $K_{u\alpha} \approx 1.65$. So

$$Q_{u\alpha} = (\frac{\pi}{2})(1)(108)(10)^2 (1.65)(\cos 40°)(\tan 33.95) = 14,436 \text{ lb}$$

$$\approx \mathbf{14.4 \ kips}$$

6.5 UPLIFT CAPACITY OF RIGID VERTICAL PILES UNDER OBLIQUE PULL

Under certain circumstances, rigid piles embedded in sand may be subjected to oblique pull as shown in Figure 6-19. The oblique pull to the pile is applied at an angle θ with the vertical. The gross ultimate uplift capacity of the pile measured *in the direction of the load application*, $Q_{u(g)\theta}$, can be given as

$$Q_{u(g)\theta} = Q_{u\theta} + W_p \cos\theta \qquad (6-37)$$

where $Q_{u\theta}$ = net ultimate uplift capacity
W_p = effective self-weight of the pile

For *rigid piles* (that is, $L/D \approx 15$ or less) the plot of net load versus pile displacement in the direction of the pull (Δ_θ) will be of the nature as shown in Figure 6-20 (Das, 1977). From this figure it can be shown that, for vertical pull (that is, $\theta = 0$), the net load gradually increases with vertical displacement up to a maximum value ($Q_{u\theta}$) at which complete pullout of the pile occurs. However, for $\theta > 0$, the net load increases with Δ_θ rather rapidly up to a certain value beyond which the load-displacement plot becomes practically

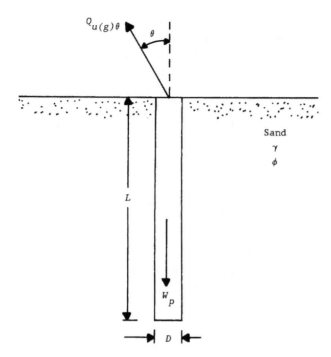

Figure 6-19   Rigid vertical pile subjected to inclined pull

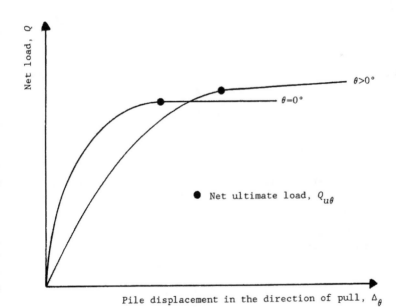

Figure 6-20   Nature of net load versus pile displacement for rigid pile subjected to oblique pull

linear. The point at which the load-displacement plot becomes practically linear is defined as the net ultimate load ($Q_{u\theta}$).

Based on laboratory model test results, Meyerhof (1973b) suggested a semi-empirical relationship to estimate the gross ultimate uplift capacity, $Q_{u(g)\theta}$, which is of the form

$$\left(\frac{Q_{u(g)\theta}\cos\theta}{Q_{u(g)V}}\right) + \left(\frac{Q_{u(g)\theta}\sin\theta}{Q_{u(g)H}}\right)^2 = 1 \tag{6-38}$$

where $Q_{u(g)V}$ = gross ultimate uplift capacity of the pile with $\theta = 0$

$Q_{u(g)H}$ = gross ultimate lateral resistance of the pile with $\theta = 90°$

It is important to realize that $Q_{u(g)V} = Q_{u(g)}$ (that is, the gross ultimate uplift capacity of the pile with $\theta = 0°$ as discussed in Sections 6.2 and 6.3). For $L/D \leq 15$, without loss of much accuracy, the following relationships may be used to estimate the magnitudes of $Q_{u(g)V}$ and $Q_{u(g)H}$.

Bored Piles with Circular Cross Section

$$Q_{u(g)V} = \underbrace{\frac{\pi}{2}D\gamma L^2 K_u \tan\delta}_{\text{Equation (6-13)}} + W_p \tag{6-39}$$

$$Q_{u(g)H} = G\cdot\gamma D^3 \tag{6-40}$$

where

$$G = \frac{Q_{u(g)H}}{\gamma D^3} \tag{6-41}$$

The variation of the nondimensional parameter G with L/D and soil friction angle ϕ is shown in Figure 6-21. These values are based on the analysis of Broms (1965).

Bored Piles with Square Cross Section

$$Q_{u(g)V} = \underbrace{2D\gamma L^2 K_u \tan\delta}_{\text{Equation (6-14)}} + W_p \tag{6-42}$$

and

$$Q_{u(g)H} = \underset{\uparrow}{G} \cdot \gamma D^3 \tag{6-40}$$

Figure 6-21

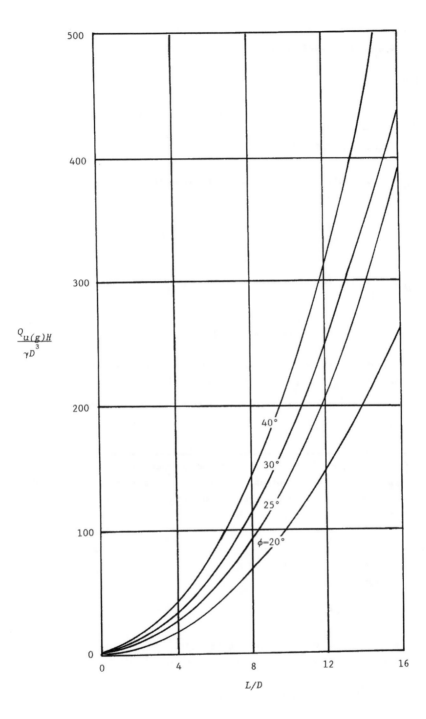

Figure 6-21   Plot of $G = Q_{u(g)H}/\gamma D^3$ based on Broms' analysis (1965)

Das, Seeley and Raghu (1977) suggested the use of Equation (6-40) for determination of $Q_{u(g)H}$ for bored piles.

Driven Piles

$$Q_{u(g)V} = \underbrace{\frac{1}{2}\gamma DL^2 K_u'}_{\text{Equation (6-22a)}} + W_p \text{ (for piles with circular cross section)} \qquad (6\text{-}43a)$$

$$Q_{u(g)V} = \underbrace{0.637\gamma DL^2 K_u'}_{\text{Equation (6-22b)}} + W_p \text{ (for piles with square cross section)} \qquad (6\text{-}43b)$$

and

$$Q_{u(g)H} = \frac{1}{2}\gamma DL^2 K_h' \text{ (for piles with circular and square cross section)} \qquad (6.44)$$

where K_h' = coefficient of lateral resistance

The variation of K_h' with soil friction angle ϕ and L/D is shown in Figure 6-22. These values of K_h' were proposed by Meyerhof (1973b).

Example

6.6

Refer to Figure 6-19. Given, for the concrete pile: length, $L = 12$ ft; diameter, $D = 12$ in. Given, for the soil: friction angle, $\phi = 35°$; relative density = 60%; unit weight = 110 lb/ft^3. Calculate $Q_{u(g)\theta}$ for $\theta = 30°$ Assume that the pile is a bored pile.

Solution

Given: $L/D = 12/1 = 12 < 15$ -- O.K.

From Equation (6-39)

$$Q_{u(g)V} = \frac{\pi}{2}D\gamma L^2 K_u \tan \delta + W_p$$

From Figure 6-5, for $\phi = 35°$, the magnitude of K_u is about 1.83. Also for $D_r = 60\%$, the value of δ/ϕ is about 0.97.

$$W_p = (\frac{\pi}{4}D^2 L)(\text{unit weight of concrete})$$

$$= [(\frac{\pi}{4})(1)^2(12)](150 \text{ lb/ft}^3) = 1413.7 \text{ lb}$$

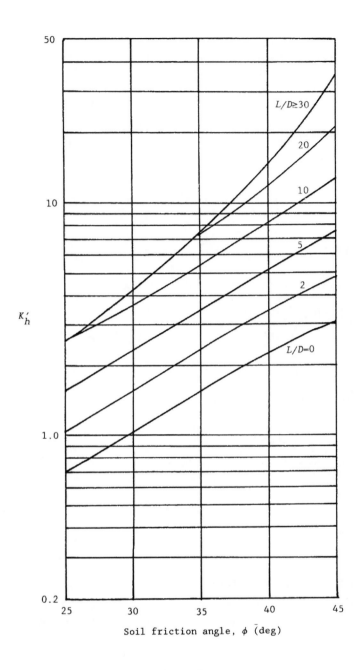

Figure 6-22 Variation of Meyerhof's K_h' with soil friction angle ϕ (1973b)

So

$$Q_{u(g)V} = (\tfrac{\pi}{2})(1)(110)(12)^2(1.83)\tan(0.97\times35) + 1413.7$$

$$\approx 32{,}068 \text{ lb} \approx 32.07 \text{ kips}$$

Again, from Equation (6-40)

$$Q_{u(g)H} = G \cdot \gamma D^3$$

For $L/D = 12$ and $\phi = 35°$, from Figure 6-21, the magnitude of G is 242. So

$$Q_{u(g)H} = (242)(110)(1)^3 = 26{,}620 \text{ lb} \approx 26.6 \text{ kips}$$

Now, from Equation (6-38)

$$\left[\frac{Q_{u(g)\theta}\cos \theta}{Q_{u(g)H}}\right] + \left[\frac{Q_{u(g)\theta}\sin \theta}{Q_{u(g)H}}\right]^2 = 1$$

or

$$\frac{[Q_{u(g)\theta}](\cos 30)}{32.07} + [Q^2_{u(g)\theta}](\frac{\sin 30}{26.6})^2 = 1$$

$$0.027Q_{u(g)\theta} + 0.00035Q^2_{u(g)\theta} = 1$$

or

$$Q^2_{u(g)\theta} + 77.14Q_{u(g)\theta} \quad 2857.14 = 0$$

$$Q_{u(g)\theta} \approx \textbf{27 kips}$$

6.6   UPLIFT CAPACITY OF GROUP PILES

Foundations subjected to uplifting loads may sometimes be constructed on a group pile (Figure 6-23). For the group pile shown in Figure 6-23, the length and diameter of all piles are L and D, respectively. All piles are placed at a center-to-center spacing of s. The number of rows and columns in the group plan are m and n, respectively. The gross and net ultimate uplift capacities of the group pile are related as

$$Q_{ug(g)} = Q_{ug} + W_{gp} \tag{6-45}$$

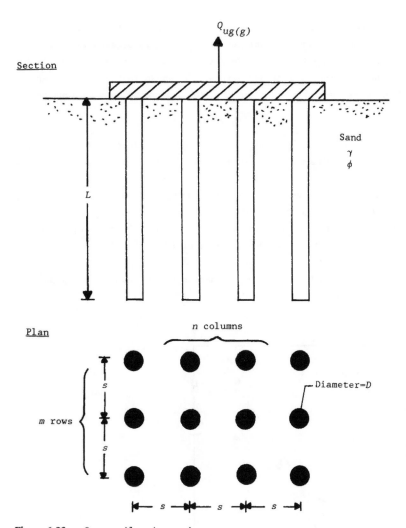

Figure 6-23   Group piles in sand

where $Q_{ug(g)}$ = gross ultimate uplift capacity of the pile group
$\quad\quad\quad Q_{ug}$ = net ultimate uplift capacity of the pile group
$\quad\quad\quad W_{gp}$ = effective self-weight of the piles in the group and the pile
$\quad\quad\quad\quad$ cap

In the conventional sense, the group efficiency η can be defined as

$$\eta = \frac{Q_{ug}}{m \cdot n \cdot Q_u} \tag{6-46}$$

222

In order to evaluate the parameters that affect the group efficiency of piles, Das (1989) provided a number of model test results in loose sand with $L/D = 15$ and 20. For these tests the relative density of compaction was kept at 47.6%. The variation of the group efficiency with s/D is shown in Figures 6-24 and 6-25. Based on these model test results, the following conclusions can be drawn:

1. For a given soil, D_r, and number of piles in a group, the group efficiency increases almost linearly with s/D up to a maximum value of 100%.
2. For a given soil and s/D, the group efficiency decreases with the increase of the number of piles in a group.
3. For a given soil, s/D, and number of piles in a group, the efficiency decreases with the increase of L/D.

More laboratory and field test results are necessary to quantify the group efficiency and develop a parametric relationship.

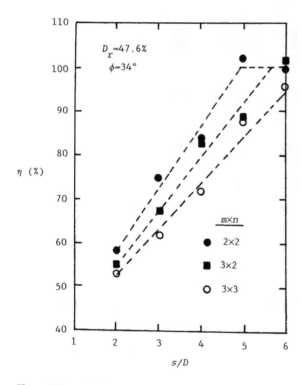

Figure 6-24   Model test results of Das (1989)--variation of η versus s/D (for $L/D = 15$)

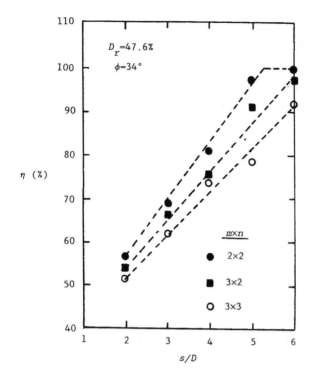

Figure 6-25   Model test results of Das (1989)--variation of η versus s/D (for $L/D = 20$)

6.7   FACTOR OF SAFETY

In all cases it is recommended that a factor of safety, F_s, of at least 2 be used to obtain the net uplift capacity, that is

$$\text{Net allowable uplift capacity} = \frac{\text{net ultimate uplift capacity}}{F_s}$$

For group piles with conventional spacing of $s = 3D$ to $4D$, the net allowable uplift capacity can be assumed as

$$Q_{ag} = \frac{m \cdot n \cdot Q_u}{F_s}$$

where Q_{ag} = net allowable uplift capacity
F_s = factor of safety $\approx$ 2 to 2.5

PILES IN CLAY ($\phi = 0$ CONDITION)

6.8 VERTICAL PILES SUBJECTED TO AXIAL PULL

Figure 6-26 shows a vertical pile embedded in a saturated or near saturated clay having an undrained cohesion of c_u. The pile is being subjected to an uplifting force. As in Equation (6-1), the gross and net ultimate uplift capacity can be expressed as

$$Q_{u(g)} = Q_u + W_p$$

For this case, however, the net ultimate uplift capacity is a function of the undrained cohesion, c_u; the pile length, L; and the perimeter of the pile cross section. Or

$$Q_u = pLc_a \qquad (6\text{-}47)$$

where p = perimeter of pile cross section

c_a = adhesion at the pile-clay interface

The adhesion is a function of the undrained cohesion. Thus

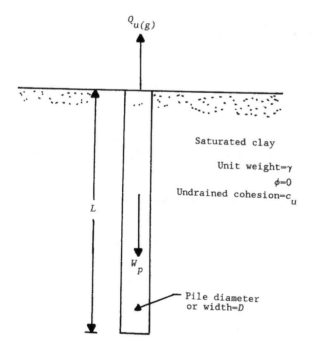

Figure 6-26   Pile embedded in saturated clay

$$c_a = f(c_u)$$

or

$$c_a = \beta c_u \qquad (6\text{-}48)$$

where β = nondimensional adhesion factor

Thus, combining Equations (6-1), (6-47) and (6-48)

$$Q_{u(g)} = \pi DL\beta c_u + W_p \quad \text{(for circular piles)} \qquad (6\text{-}49)$$

and

$$Q_{u(g)} = 4DL\beta c_u + W_p \quad \text{(for square piles)} \qquad (6\text{-}50)$$

The important parameter in the preceding two equations is the adhesion factor β which needs to be determined for estimation of the ultimate uplift capacity. Following is a summary of available published results.

Cast *In situ* Piles

A number of field test results for the ultimate uplift capacity of cast *in situ* concrete piles were reported by Patterson and Urie (1964), Turner (1962), Mohan and Chandra (1961) and Sowa (1970). Based on these field test results, the adhesion factors have been calculated and are shown in Figure 6.27 The average plot of the variation of β with c_u can be expressed as (Das and Seeley, 1982)

$$\beta = 0.9 \quad 0.00625c_u \geq 0.4 \qquad (6\text{-}51)$$

where c_u = undrained cohesion in kN/m^2

For a given value of c_u, the magnitude of β determined from Equation (6-51) is slightly lower than that recommended by Tomlinson (1957) for calculation of the skin resistance of concrete piles under compressive loading.

Metal Piles

Das and Seeley (1982) reported several laboratory model test results for the ultimate uplift capacity of metal piles in saturated clay. Based on their observations it appears that, for metal piles

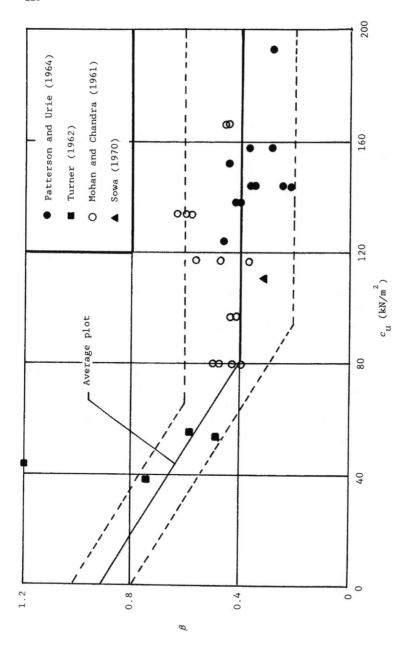

Figure 6-27 Variation of β versus c_u for cast-in-situ concrete piles

$$\beta = 0.715 \quad 0.0191c_u \geq 0.2 \qquad (6\text{-}52)$$
$$\uparrow$$
$$(kN/m^2)$$

Figure 6-28 shows a comparison between Equations (6-51) and (6-52). It can be seen that, for all values of c_u, the β factor for metal piles is lower than that for the cast *in situ* concrete piles.

6.9 LOAD-DISPLACEMENT RELATIONSHIP FOR VERTICAL PILES SUBJECTED TO AXIAL UPLIFT

Das and Seeley (1982) reported that, for metal piles with $L/D \leq 16$, the net ultimate uplift capacity is realized when the pile undergoes an axial uplift of about $0.05D$ ($\approx \Delta_u$). Based on their model test results, they suggested a nondimensional relationship between the net load Q and vertical displacement Δ which is of the form

$$\overline{Q} = \frac{\overline{\Delta}}{a + b\overline{\Delta}} \qquad (6\text{-}53)$$

where $\overline{Q} = \dfrac{Q}{Q_u}$ \qquad (6-54)

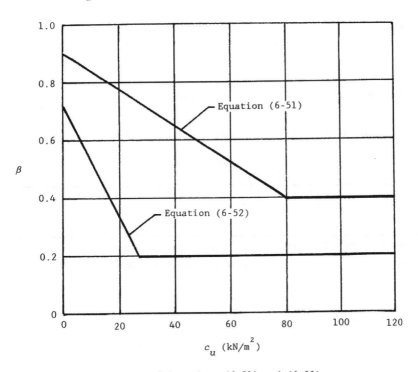

Figure 6-28　Comparison of Equations (6-51) and (6-52)

$$\overline{\Delta} = \frac{\Delta}{\Delta_u} \tag{6-55}$$

Q = net load at an axial displacement Δ

Q_u = net ultimate load at an axial displacement Δ_u

a,b = constants

Equation (6-53) can be used to make preliminary estimation of the axial displacement of a pile for a net allowable load Q. The average values of the constants a and b can be taken as 0.2 and 0.8, respectively.

Example
6.7

A vertical concrete pile having a square cross section of 0.3 m × 0.3 m and a length of 8 m is embedded in a saturated clay having an undrained cohesion of 60 kN/m^2 Estimate the net ultimate uplift capacity.

Solution

From Equation (6-50)

$$Q_u = 4DL\beta c_u$$

$D = 0.3$ m; $L = 8$ m; $c_u = 60$ kN/m^2 From Equation (6-51)

$$\beta = 0.9 \quad 0.00625c_u = 0.9 \quad (0.00625)(60) = 0.525$$

So

$$Q_u = (4)(0.3)(8)(0.525)(60) = \textbf{302.4 kN}$$

Example
6.8

Refer to Example Problem 6.7. For an allowable net uplift load of 100 kN, estimate the vertical displacement of the pile.

Solution

From Equation (6-54)

$$\overline{Q} = \frac{Q}{Q_u} = \frac{100}{302.4} = 0.331$$

Again, from Equation (6-53)

$$\bar{Q} = \frac{\bar{\Delta}}{a + b\bar{\Delta}}$$

$a \approx 0.2; \ b \approx 0.8.$ So

$$0.331 = \frac{\bar{\Delta}}{0.2 + 0.8\bar{\Delta}}$$

Or

$$0.662 + 0.2648\bar{\Delta} = \bar{\Delta}$$

$$\bar{\Delta} = \frac{0.662}{0.735} = 0.9$$

However

$$\bar{\Delta} = \frac{\Delta}{\Delta_u}$$

So

$$\Delta = (\bar{\Delta})(\Delta_u) = (0.9)(0.05D) = (0.9)(0.05)(0.3) = 0.0135 \ m = \textbf{13.5 mm}$$

6.10   INCLINED PILE SUBJECTED TO AXIAL PULL

Figure 6-29 shows an inclined pile subjected to axial pull. The inclination of the pile with respect to the vertical is equal to α. For this condition, the gross uplift capacity can be given by Equation (6-23), or

$$Q_{u(g)\alpha} = Q_{u\alpha} + W_p\cos \alpha$$

The magnitude of the net ultimate uplift capacity, $Q_{u\alpha}$, can be given as

$$Q_{u\alpha} = pL\beta c_u \tag{6-56}$$

where p = perimeter of the pile = $\begin{cases} \pi D \text{ for circular piles} \\ 4D \text{ for square piles} \end{cases}$

The magnitude of the adhesion factor can be estimated from Equations (6-51) or (6-52) depending on the pile type.

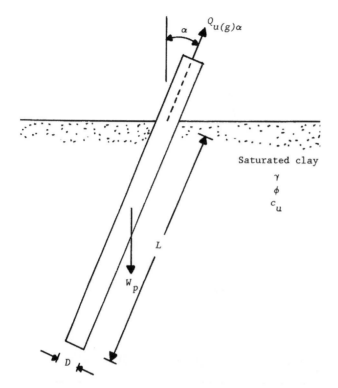

Figure 6-29   Inclined pile embedded in clay subjected to axial pull

6.11   UPLIFT CAPACITY OF VERTICAL PILE SUBJECTED TO INCLINED PULL

As in Section 6.5, Figure 6-30 shows a vertical pile embedded in saturated clay which is subjected to an inclined pull. The ultimate uplift capacity of the pile measured in the direction of the pull can be given by Equation (6-37), or

$$Q_{u(g)\theta} = Q_{u\theta} + W_p\cos\theta$$

For rigid piles (that is, $L/D \leq 15$ to 20), the gross ultimate uplift capacity $Q_{u(g)\theta}$ can also be given by Equation (6-38) (Meyerhof, 1973b). So

$$\left[\frac{Q_{u(g)\theta}\cos\theta}{Q_{u(g)V}}\right] + \left[\frac{Q_{u(g)\theta}\sin\theta}{Q_{u(g)H}}\right] = 1$$

The magnitude of $Q_{u(g)V}$ can be estimated from Equations (6-49) or (6-50), or

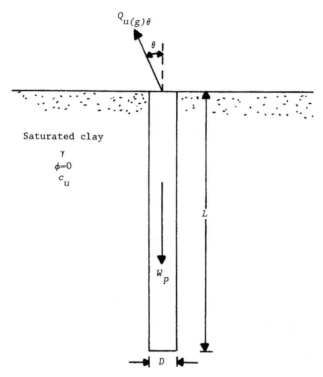

Figure 6-30   Vertical pile embedded in saturated clay subjected to inclined pull

$$Q_{u(g)V} = pL\beta c_u + W_p \qquad\qquad (6-57)$$

where $p = \begin{cases} \pi D \text{ for circular piles} \\ 4D \text{ for square piles} \end{cases}$

The gross ultimate lateral load $Q_{u(g)H}$ can be estimated as (Meyerhof, 1973b)

$$Q_{u(g)H} = c_u K_h'' LD \qquad\qquad (6-58)$$

where K_h'' = coefficient of lateral resistance

Based on experimental results, Meyerhof (1973b) proposed that

$$K_h'' = 1 + 0.8(\tfrac{L}{D}) \le 3 \qquad\qquad (6-59)$$

6.12   UPLIFT CAPACITY OF GROUP PILES IN CLAY

Research relating to the uplift capacity of group piles (Figure 6-23) is rather scarce at the present time. The gross and net ultimate uplift capacities of group piles can be related by Equation (6-45), or

232

$$Q_{ug(g)} = Q_{ug} + W_{gp}$$

Also the group efficiency can be expressed by Equation (6-46)

$$\eta = \frac{Q_{ug}}{m \cdot n \cdot Q_u}$$

Das (1990) provided several model test results for group efficiency of piles in soft clay with L/D = 15 and 20 and c_u = 10.06 kN/m^2 and 22.5 kN/m^2, and these results are shown in Figures 6-31, 6-32, 6-33 and 6-34. From these figures, the following conclusions can be drawn:

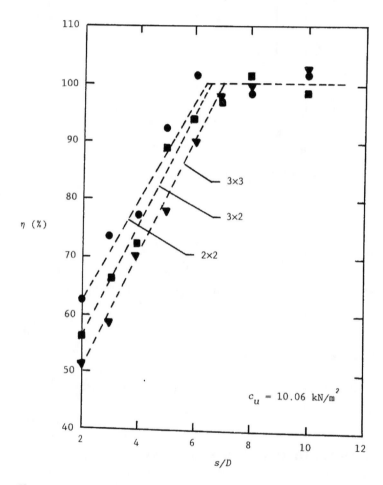

Figure 6-31   Model test results of Das (1990)--variation of η versus s/D (L/D = 15; c_u = 10.06 kN/m^2)

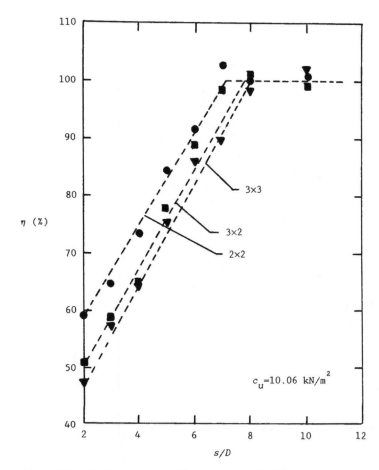

Figure 6-32   Model test results of Das (1990)--variation of η
versus s/D ($L/D = 20$; c_u = 10.06 kN/m²)

1.   For a given clay (that is, c_u), L/D, and number of piles in a group, the magnitude of η increases linearly with s/D.

2.   For a given s/D ratio and clay, the group efficiency decreases with the increase of the number of piles ($m \times n$) in the group.

3.   For a given clay soil (that is, c_u) and number of piles in the group, the efficiency decreases with the increase of the L/D ratio.

4.   For a given L/D ratio, number of piles in a group, and s/D, the increase of c_u results in a decrease of the magnitude of the group efficiency.

5.   For short piles (that is, $L/D \leq 20$), it appears that

$$(\frac{s}{D})_{\eta=100\%} \approx 0.4 \text{ to } 0.5(\frac{L}{D})$$

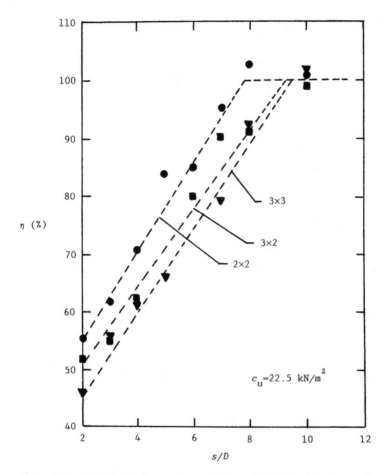

Figure 6-33   Model test results of Das (1990)--variation of η
versus s/D (L/D = 15; c_u = 22.5 kN/m^2)

REFERENCES

Broms, B.B., 1965. Design of laterally loaded piles. *J. Soil Mech Found.
Div.*, ASCE, 91(3):79-97.

Das, B.M., 1990. Group efficiency of metal piles in clay subjected to uplifting load. *Proc.*, OMAE Conf., ASME, Houston, TX.

Das, B.M., 1989. Ultimate uplift capacity of piles and pile groups in
granular soil. *Proc.*, Intl. Conf. on Piling and Deep Foundations,
London, DFI:241-246.

Das, B.M., 1983. A procedure for estimation of uplift capacity of rough
piles. *Soils and Found.*, Japan, 23(3):122-126.

Das, B.M. and Seeley, G.R., 1982. Uplift capacity of pipe piles in saturated
clay. *Soils and Found.*, Japan, 22:91-94.

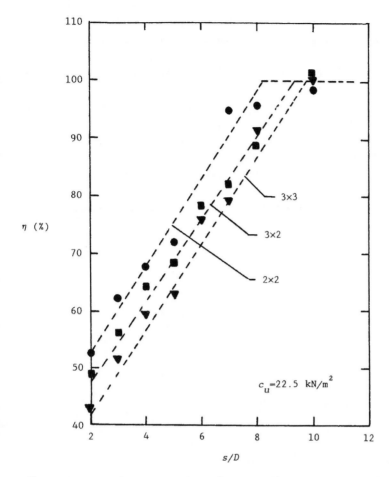

Figure 6-34   Model test results of Das (1990)--variation of η
versus s/D (L/D = 20; c_u = 22.5 kN/m^2)

Das, B.M., Seeley, G.R. and Raghu, D., 1977. Uplift capacity of model piles
under oblique loads. *J. Geotech. Engr. Div.*, ASCE, 1202(9):1009-1013.

Das, B.M., Seeley, G.R. and Pfeifle, T.W., 1977. Pullout resistance of rough
rigid piles in granular soils. *Soils and Found.*, Japan, 17(3):72-77.

Meyerhof, G.G., 1973a. Uplift resistance of inclined anchors and piles.
Proc., VIII Intl. Conf. Soil Mech. Found. Engrg., Moscow, USSR, 2.1:167-172.

Meyerhof, G.G., 1973b. The uplift capacity of foundations under oblique
loads. *Can. Geotech. J.*, 10(1):64-70.

Mohan, D. and Chandra, S., 1961. Frictional resistance of bored piles in ex-
pansive clays. *Geotechnique*, 11(4):294-301.

Patterson, G. and Urie, R.I., 1964. Uplift resistance of full size tower foundations. *Proc.*, Conf. Internationale des Grande Reseaux Electrique a haute Tension, Paper No. 203, Paris.

Sowa, V.A., 1970. Pulling capacity of concrete cast-*in situ* piles. *Can. Geotech. J.*, 7:482-493.

Tomlinson, M.J., 1957. The adhesion of piles driven in clay soils. *Proc.*, IV Intl. Conf. Soil Mech. Found. Engrg., London, 2:66-71.

Tran-Vo-Nhiem, 1971. Ultimate uplift capacity of anchor piles. *Proc.*, IV Conf. Soil Mech., Budapest:829-836.

Turner, E.A., 1962. Uplift resistance of transmission tower footing. *J. Power Div.*, ASCE, 88(2):17-33.

Vesic, A.S., 1970. Tests on instrumented piles, Ogeechee River site. *J. Soil Mech. Found. Div.*, ASCE, 96(2):561-584.

INDEX